HIGHER ARITHMETIC,

DESIGNED FOR THE USE OF

HIGH SCHOOLS, ACADEMIES, AND COLLEGES;

IN WHICH

SOME ENTIRELY NEW PRINCIPLES ARE DEVELOPED, AND MANY CONCISE AND EASY RULES GIVEN, WHICH HAVE NEVER BEFORE APPEARED IN ANY ARITHMETIC:

WITH AN APPENDIX.

By GEORGE R. PERKINS, A. M.,

PRINCIPAL AND PROFESSOR OF MATHEMATICS OF THE NORMAL SCHOOL OF THE STATE OF NEW YORK, AUTHOR OF ELEMENTARY ARITHMETIC, ELEMENTARY ALGEBRA, TREATISE ON ALGEBRA, ELEMENTS OF GEOMETRY, ETC., ETC.

STEREOTYPE EDITION, REVISED AND IMPROVED.

UTICA:
PUBLISHED BY H. H. HAWLEY & CO

HARTFORD:
J. H. MATHER & CO.
1849.

STEREOTYPED BY
RICHARD H. HOBBS,
HARTFORD, CONN.

PRINTED BY
CASE, TIFFANY & CO.,
HARTFORD, CONN.

PREFACE

TO THE SECOND EDITION.

IN preparing this edition, the whole work has been carefully revised, and many of the subjects re-written and modified, with a view to make them more clearly comprehended by the pupil.

Having explained fully, in the ELEMENTARY ARITHMETIC, the subjects of Duodecimals and Alligation, we have omitted them in the present edition. We have excluded the chapter on Permutations, Combinations, and Variations, to make room for what we deemed matter of more importance. We have also placed the chapters on the Progressions after Evolution, because several cases under the Progressions are wrought by the aid of Evolution.

Under the first chapter, we have added some new things concerning Prime Numbers, and have explained the use of ERATOSTHENES' SIEVE in finding these numbers. After Continued Fractions, we have added LAMBERT'S Method of De-compounding Vulgar Fractions. We have given a new rule for the Equation of Payments, under the subject of Discount. This rule is deduced from considering the *equivalent present values*. By this method, we discover that the usual rule is erroneous, if we consider the present value of the several debts.

We have added the answers to all the questions; and, where we thought the operations too difficult, we have either given the whole work, or so much as to leave nothing to perplex the beginner.

In all these modifications we have been guided, not only by our own practical experience in using the book, but by the many kind suggestions which have been received from distinguished teachers.

GEO. R. PERKINS.

UTICA, *September*, 1844.

PREFACE

TO THE FOURTH EDITION.

THE present edition, which is *stereotyped*, has been carefully revised, and, in some cases, new matter added in the body of the work, without, however, changing the order of the different articles. But what most distinguishes it from the preceding ones, is the addition of an APPENDIX.

In this Appendix, we have discussed pretty fully what may properly be considered the philosophy of some of the more difficult operations, as well as interesting properties of numbers. It is believed that much here added will be found instructive and interesting to the lover of arithmetical operations.

GEO. R. PERKINS.

UTICA, *April*, 1848.

CONTENTS.

CHAPTER I.

CHAPTER II.

CHAPTER III.

CHAPTER IV.

CHAPTER V.

CHAPTER VI.

CHAPTER VII.

CHAPTER VIII.

CHAPTER IX.

ARITHMETIC.

CHAPTER I.

DEFINITIONS.

1. Any whole number is called an *integer*.

2. Any number which can be divided by 2, without a remainder, is called an *even number*.

3. All numbers which can not be divided by 2, without a remainder, are called *odd numbers*.

4. Any number which can be produced by multiplying two or more numbers together, each of which is greater than a unit, is called a *composite number*. Thus, 35 is a composite number, since it can be produced by multiplying 5 and 7 together.

5. The numbers which are multiplied together to produce a composite number, are called *factors*. Thus, 3 and 8 are factors of 24; so, also, are 4 and 6.

6. A composite number which is composed of two equal factors, is called a *square number*. Thus, 4, 9, 16, and 49, are square numbers.

7. A composite number which is composed of three equal factors, is called a *cube number*. Thus, 8, 27, and 64, are cube numbers.

8. One of the equal factors which compose a square number, is called the *square root* of the number. Thus, 7 is called the square root of 49.

9. One of the equal factors which compose a cube

number, is called the *cube root* of the number. Thus, 3 is the cube root of 27.

10. All numbers which are not composite, are called *prime numbers.* Thus, 1, 2, 3, 5, 7, 11, and 13, are prime numbers.

11. Unity divided by a number is the *reciprocal* of that number. Thus, $\frac{1}{2}$, $\frac{1}{3}$, and $\frac{1}{4}$, are the reciprocals of 2, 3, and 4.

12. A number taken without regard to the particular kind of unit, is called an *abstract number.*

13. A number considered in reference to a particular unit, is called a *concrete number*, or a *denominate number.* Thus, 8 yards is a denominate number, whose unit is one yard; 37 days is a denominate number, whose unit is one day.

SYMBOLS.

2. THE symbol =, is called the sign of *equality;* and denotes that the quantities between which it is placed are equal or equivalent to each other. Thus, \$1=100 cents; which is read, one dollar equals one hundred cents.

2. The symbol +, is called *plus;* and denotes that the quantities between which it is placed are to be added together. Thus, $6+2=8$; which is read, six and two added equals eight.

3. The symbol −, is called *minus*, and denotes that the quantity which is placed at the right of it is to be subtracted from the quantity on the left. Thus, $6-2=4$; which is read, six diminished by two equals four.

4. The symbol $\times$, is called the sign of *multiplication*, and denotes that the quantities between which it is placed are to be multiplied together. Thus, $6\times2=12$; which is read, six multiplied by two equals twelve.

Multiplication is sometimes expressed by a dot, (.); Thus, $3 . 4$, is the same as 3×4.

5. The symbol $\div$, is called the sign of *division*, and denotes that the quantity on the left of it is to be divided by the quantity on the right. Thus, $6\div2=3$; which is read, six divided by two equals three.

Division is also denoted by placing the divisor under the dividend, with a horizontal line between them, like a vulgar fraction. Thus, $\frac{6}{2}$ is the same as $6\div2$.

6. A number placed above another number, a little to the right, is called an *exponent*. Thus, in the expressions, 6^2, 7^3, ^{2} and ^{3} are exponents of 6 and 7, respectively.

7. An exponent placed over a quantity, denotes that the quantity is to be used as a factor as many times as there are units in the exponent. Thus, $2^4=2\times2\times2\times2=16$.

8. When the exponent is ^{2}, the result is called the *second power* of the quantity over which it is placed. Thus, $7^2=7\times7=49=$ the second power of 7.

9. When the exponent is ^{3}, the result is called the *third power* of the quantity over which it is placed. Thus, $4^3=4\times4\times4=64=$ the third power of 4.

The higher powers are denoted in the same way.

10. The symbol $\sqrt{}$, denotes that the *square root* of the quantity over which it is placed is to be taken. Thus, $\sqrt{4}=2$; which is read, the square root of four equals two.

11. The symbol $\sqrt[3]{}$, denotes, in a similar manner, the *cube root* of the number over which it is placed. Thus, $\sqrt[3]{64}=4$; which is read, the cube root of sixty-four equals four.

The roots of higher dimensions are denoted in a similar way.

12. The symbol $\therefore$ is equivalent to the phrase *therefore* or *consequently*. Thus $6^2=36$, and $4\times9=36\therefore 6^2=4\times9$; which is read, the square of six equals thirty-six, and the product of four and nine equals thirty-six; therefore, the square of six equals the product of four and nine.

13. The parenthesis, ( ), when it incloses several quantities, requires these quantities to be regarded as one single quantity. Thus, $(5+3)\times7=56$; which is read, the sum of five and three multiplied by seven equals fifty-six.

EXAMPLES

ILLUSTRATING THE FOREGOING DEFINITIONS AND SYMBOLS.

3. THE expression, $11+5-2=2\times7=28\div2=14$, when translated into common language, becomes the sum of eleven and five diminished by two, equals the product of two and seven, equals twenty-eight divided by two, equals fourteen.

2. The expression, $\frac{42-8}{2}+3=4\times5=20$, is equivalent to the following:—forty-two diminished by eight, and the remainder divided by two, and the quotient increased by three, equals four multiplied into five, equals twenty.

3. The expression, $\sqrt{144}=3\times4=36\div3=12$, is the same as the square root of one hundred and forty-four, equals three multiplied into four, equals thirty-six divided by three, equals twelve.

4. Translate the expression, $(10+3)\times7=182\div2=91$, into common language.

Ans. The sum of ten and three multiplied by seven is equal to one hundred and eighty-two divided by two, which is equal to ninety-one.

5. Translate the expression, $(\sqrt{16}+7)\times4=\sqrt[3]{64}\times11$, into common language.

Ans. The square root of sixteen increased by seven, and the sum multiplied by four, is equal to the cube root of sixty-four multiplied by eleven.

6. Translate the expression, $(\sqrt{49}-\sqrt[3]{64})\times3=20-11$, into common language.

Ans. The square root of forty-nine diminished by the cube root of sixty-four, and the difference multiplied by three, is equal to twenty diminished by eleven.

7. What expression is equivalent to the following: "Five times nine divided by three, and that quotient multiplied by seven, equals the square of ten increased by five?"

Ans. $\frac{5\times9}{3}\times7=10^2+5.$

8. What expression is equivalent to the following. "Three times twenty-one, increased by five times seven, and diminished by three times the square of four, is equal to twice the square of five?"

Ans. $3\times21+5\times7-3\times4^2=2\times5^2.$

9. What expression is equivalent to the following: "The cube root of sixty-four, increased by two, and

the sum multiplied by ten, is equal to the square of eight diminished by four?"

Ans. $(\sqrt[3]{64}+2)\times 10=8^2-4$.

MULTIPLICATION OF COMPOUND EXPRESSIONS.

4. In multiplication, the multiplier must always be regarded as an abstract number.

The multiplicand may be a quantity or denominate number of any kind. Thus, we may repeat $7 once, twice, thrice, or any number of times, but we could not multiply dollars by yards, pounds, hours, or by any other denominate number.

It is sometimes proposed to multiply money by money, as 2 s. 6 d. by 2 s. 6 d., but this is not philosophically correct. We may repeat 2 s. 6 d. once, twice, &c., but we cannot repeat it 2 s. 6 d. times.

In estimating the cost of 20 bushels of apples at 25 cents per bushel, we do not repeat 25 cents 20 bushels of times, neither do we repeat 20 bushels 25 cents times; but, having fixed upon a quantity of apples equal to one bushel as our unit, we find that the quantity whose cost is to be estimated is 20 bushels, or twenty times the unit; now as one bushel is worth 25 cents, 20 bushels, being 20 times as many apples, must be worth 20 times as many cents as one bushel; we therefore repeat 25 cents twenty times, not 20 bushels of times, and thus obtain 500 cents, or 5 dollars, for the whole cost. And, in a similar way, we might show that, in all operations of multiplication, the multiplier is an abstract number, denoting how many *times* the multiplier is to be repeated. The product must evidently be of the same name as the multiplicand, since repeating a quantity once, twice, thrice, or any number of times, cannot change its denomination. Sometimes the multiplicand is also an abstract number.

It would be equally absurd to suppose the multiplier to be a *negative* quantity; for we could not repeat a quantity a *minus* number of times. If, then, in the course of an operation, we have for factors a positive and a negative quantity, we must regard the positive factor as the multiplier, and the negative quantity as the multiplicand; as, for example, if we wish the product of 4 and -7, we must repeat -7 four times, and the result will still be negative, we shall thus obtain -28. Or we might have multiplied the 4 by 7, and changed the sign of the product. From which we see that when a minus quantity occurs in the multiplier, we are to multiply by it considered as positive, and then to change the sign of the product.

Applying this principle to the case when both factors are negative, as, for example, -5 multiplied by -7. In this case, it will be necessary to repeat -5 seven times, and then to change the sign of the product, we thus obtain for our result, 35. From what has been said and done, we have for the multiplication of compound expressions the following

RULE.

Multiply each term of one of the factors by each term of the other factor, observing that like signs produce plus, and unlike signs produce minus.

Let it be required to multiply $3+2$ by $4+5$.

We must repeat $3+2$ as many times as there are units in $4+5$.

First, repeating $3+2$ as many times as there are units in 4, we get $(3+2)\times 4=12+8$, for first partial product:

Secondly, repeating $3+2$ as many times as there are units in 5, we get $(3+2)\times 5=15+10$, for second partial product:

Hence, $3+2$, repeated as many times as there are units in $4+5$, becomes $(3+2)\times(4+5)=12+8+15+10$.

Again, let it be required to multiply $7-3$ by $4+2$: Proceeding as in the last example, we find $(7-3)\times(4+2)=28-12+14-6$.

In a similar way, we find that $4-3$, multiplied by $3-2$, gives $(4-3)\times(3-2)=12-9-8+6$.

EXAMPLES.

1. What is the product of $8+3$ by $6+4$?
   Ans. $48+18+32+12$.
2. What is the product of $6-2$ by $4+3$?
   Ans. $24-8+18-6$.
3. What is the product of $11-3$ by $13-7$?
   Ans. $143-39-77+21$.
4. What is the product of $3+2-1$ by $4-1+5$?
   Ans. $12+8-4-3-2+1+15+10-5$.
5. What is the product of $1+2-3$ by $4-5+6$?
   Ans. $4+8-12-5-10+15+6+12-18$.
6. What is the product of $7-9$ by $5-11$?
   Ans. $35-45-77+99$.
7. What is the product of $21-3$ by $9-2$?
   Ans. $189-27-42+6$.
8. What is the product of $1+7+5$ by $2+3$?
   Ans. $2+14+10+3+21+15$.

SINGULAR PROPERTY OF THE FIGURE 9.

5. *Every number will divide by* 9, *when the sum of its digits is divisible by* 9.

For, take any number, as 78534; this number is, by the nature of decimal arithmetic, the same as 70000+8000+500+30+4.

$$\begin{aligned} \text{Now, } 70000 &= 9999\times 7+7 \\ 8000 &= 999\times 8+8 \\ 500 &= 99\times 5+5 \\ 30 &= 9\times 3+3 \\ 4 &= \quad\ +4 \end{aligned}$$

$\therefore$ 78534=9999×7+999×8+99×5+9×3+(7+8+5+3+4.)

Now, since each expression, 9999×7, 999×8, 99×5, and 9×3, is divisible by 9, it follows that the first number, 78534, will be divisible by 9 when the sum of its digits (7+8+5+3+4) is.

Hence, it follows that any number being diminished by the sum of its digits, will become divisible by 9.

Also, any number divided by 9, will leave the same remainder as the sum of its digits when divided by 9.

The above properties belong to the digit 3, as well as to that of 9, since 3 is a divisor of 9. No other digit has such properties.

NOTE.—These singular properties of the digit 9, have been made use of by many authors for proving the work of the four fundamental rules of arithmetic.

PRIME NUMBERS.

6. No even number can, with the single exception of the number 2, be a prime, since all even numbers are divisible by 2. It is also evident that there are many odd numbers which are not primes. If we write in order the natural series of odd numbers, we discover that every third term, counting from 3, is divisible by 3; every fifth term, counting from 5, is divisible by 5; every seventh term, counting from 7, is divisible by 7, and so on.

Commencing at 3, under every third term, I have placed a small figure, $_3$, to denote that the term under which it is placed is divisible by 3. Under every fifth term, counting from 5, I have, in like manner, placed a small $_5$, indicating that the corresponding term is divisible by 5. I have proceeded in the same way for the higher primes. Now it is evident that all the terms, under which there are no small figures found, are primes.

We may also remark, that the numbers expressed by the small figures are the different prime factors of the numbers under which they are placed.

1, 3, 5, 7, 9_3, 11, 13, $15_{3.5}$, 17, 19, $21_{3.7}$, 23, 25_5, 27_3, 29, 31, $33_{3.11}$, $35_{5.7}$, 37, $39_{3.13}$, 41, 43, $45_{3.5}$, 47, 49_7, $51_{3.17}$, 53, $55_{5.11}$, $57_{3.19}$, 59, 61, $63_{3.7}$, $65_{5.13}$, 67, $69_{3.23}$, 71, 73, $75_{3.5}$, $77_{7.11}$, 79, 81_3, 83, $85_{5.17}$, $87_{3.29}$, 89, $91_{7.13}$, $93_{3.31}$, $95_{5.19}$, 97, $99_{3.11}$, &c.

In the above operation, we have found the primes only which are less than 100; but this process may be extended as far as we wish. This method of finding the successive primes was employed by *Eratosthenes*, who inscribed the series of odd numbers upon parchment, then cutting out such numbers as he found to be composite, his parchment with its holes resembled somewhat a *sieve;* hence, this method is called *Eratosthenes' Sieve.*

The number 2, although an even number, must be regarded as coming under our definition of a prime, since the only number which will divide it is itself.

TABLE OF PRIME NUMBERS.

1	157	373	607	857	1103	1399	1637	1949	2239	2531
2	163	379	613	859	1109	1409	1657	1951	2243	2539
3	167	383	617	863	1117	1423	1663	1973	2251	2543
5	173	389	619	877	1123	1427	1667	1979	2267	2549
7	179	397	631	881	1129	1429	1669	1987	2269	2551
11	181	401	641	883	1151	1433	1693	1993	2273	2557
13	191	409	643	887	1153	1439	1697	1997	2281	2579
17	193	419	647	907	1163	1447	1699	1999	2287	2591
19	197	421	653	911	1171	1451	1709	2003	2293	2593
23	199	431	659	919	1181	1453	1721	2011	2297	2609
29	211	433	661	929	1187	1459	1723	2017	2309	2617
31	223	439	673	937	1193	1471	1733	2027	2311	2621
37	227	443	677	941	1201	1481	1741	2029	2333	2633
41	229	449	683	947	1213	1483	1747	2039	2339	2647
43	233	457	691	953	1217	1487	1753	2053	2341	2657
47	239	461	701	967	1223	1489	1759	2063	2347	2659
53	241	463	709	971	1229	1493	1777	2069	2351	2663
59	251	467	719	977	1231	1499	1783	2081	2357	2671
61	257	479	727	983	1237	1511	1787	2083	2371	2677
67	263	487	733	991	1249	1523	1789	2087	2377	2683
71	269	491	739	997	1259	1531	1801	2089	2381	2687
73	271	499	743	1009	1277	1543	1811	2099	2383	2689
79	277	503	751	1013	1279	1549	1823	2111	2389	2693
83	281	509	757	1019	1283	1553	1831	2113	2393	2699
89	283	521	761	1021	1289	1559	1847	2129	2399	2707
97	293	523	769	1031	1291	1567	1861	2131	2411	2711
101	307	541	773	1033	1297	1571	1867	2137	2417	2713
103	311	547	787	1039	1301	1579	1871	2141	2423	2719
107	313	557	797	1049	1303	1583	1873	2143	2437	2729
109	317	563	809	1051	1307	1597	1877	2153	2441	2731
113	331	569	811	1061	1319	1601	1879	2161	2447	2741
127	337	571	821	1063	1321	1607	1889	2179	2459	2749
131	347	577	823	1069	1327	1609	1901	2203	2467	2753
137	349	587	827	1087	1361	1613	1907	2207	2473	2767
139	353	593	829	1091	1367	1619	1913	2213	2477	2777
149	359	599	839	1093	1373	1621	1931	2221	2503	2789
151	367	601	853	1097	1381	1627	1933	2237	2521	2791

The preceding table contains all the primes which are not greater than 2791.

All prime numbers, except 2, are odd, and, therefore, terminate with an odd digit. Any number which ends with 5, is divisible by 5; *hence it follows that all primes, except* 2 *and* 5, *must end with one of the figures*, 1, 3, 7, or 9.

When it is required to determine whether a given number is a prime, we first notice the terminating figure; if it is different from 1, 3, 7, or 9, the number is a composite; but if it terminate with one of the above digits, we must endeavor to divide it by some one of the primes, as found in the table, commencing with 3. There is no necessity of trying 2, for 2 will divide only the even numbers. If we proceed to try all the successive primes of the table until we reach a prime which is not less than the square root of the number, without finding a divisor, we may conclude with certainty that the number is a *prime*.

The reason why we need not try any primes greater than the square root of the number, is drawn from the following consideration: If a composite number is resolved into two factors, one of which is less than the square root of the number, the other must be greater than the square root.

The square of the last prime given in our table is 7789681; hence, this table is sufficiently extended to enable us to determine whether any number not exceeding 7789681 is a prime. It is obvious that numbers may be proposed which would require by this method very great labor to determine whether they are primes, still this is the only sure and general method as yet discovered.

Tables have been calculated, giving not only all the primes up to 3036000, but also the least prime factor of the composite numbers up to the same extent.

Our table is of sufficient extent to enable us to work all ordinary examples.

Any prime number, except 2 *and* 3, *when divided by* 6, *must have a remainder of* 1 *or* 5; for all prime numbers are odd, and any odd number when divided by an even number, must leave an odd number for remainder. Hence, any odd number divided by 6, must give 1, 3, or 5, for a remainder; if the remainder is 3, the number must have been divisible by 3, since the divisor and remainder are each divisible by 3. Hence, the remainder found by dividing a prime by 6, is 1 or 5. *Therefore, by either adding one or subtracting one from any prime number greater than* 3, *it becomes divisible by* 6.

---

7. *Every number is either a prime number, or composed of prime factors.*

For, all numbers which are not prime are composite, and can, therefore, be separated into two or more factors; and, if these factors are not prime, they can again be separated into other factors, and thus the decomposition can be continued until all the factors are prime.

Hence, to resolve any composite number into its prime factors, we have this

RULE.

Divide the number by any prime number which will divide it without any remainder; then divide the quotient in the same way, and so continue until a quotient is

obtained which is a prime. Then will the successive divisors, together with the last quotient, be the prime factors required.

EXAMPLES.

1. Resolve 728 into its prime factors.

Operation.

$$\begin{array}{r|r} 2 & 728 \\ \hline 2 & 364 \\ \hline 2 & 182 \\ \hline 7 & 91 \\ \hline & 13 \end{array}$$

Therefore, $2\times 2\times 2\times 7\times 13=2^3\times 7\times 13$, are the prime factors of 728.

2. Resolve 812 into its prime factors.

Ans. $2^2\times 7\times 29$.

3. What are the prime factors of 978?

Ans. $2\times 3\times 163$.

4. What are the prime factors of 1011?

Ans. 3×337.

5. What are the prime factors of 100?

Ans. $2^2\times 5^2$.

6. What are the prime factors of 8975?

Ans. $5^2\times 359$.

7. What are the prime factors of 808?

Ans. $2^3\times 101$.

8. What are the prime factors of 707?

Ans. 7×101.

9. What are the prime factors of 1118?

Ans. $2\times 13\times 43$.

8. Suppose we wish to know whether the numbers 204 and 468 have a common factor, we proceed as follows: We decompose them into their prime factors, and thus obtain $204=2^2\times 3\times 17$, and $468=2^2\times 3^2\times 13$. Here we see that $2^2\times 3$ is common to both the numbers 204 and 468.

Hence, to find the greatest factor which is common to two or more numbers, or, as generally expressed, to find the greatest common measure of two or more numbers, we have this

RULE.

Resolve the numbers into their prime factors, (by Rule under Art. **7.**) *Then select such of the primes as are common to all the numbers, multiply them together, and the product will be the greatest common measure.*

EXAMPLES.

1. What is the greatest common measure of 1326, 3094, and 4420?

These numbers, when resolved into their prime factors, become

$$1326=2\times 3\times 13\times 17$$
$$3094=2\times 7\times 13\times 17$$
$$4420=2^2\times 5\times 13\times 17$$

The factors which are common, are 2, 13, and 17; therefore, the greatest common measure is $2\times 13\times 17=442$.

2. What is the greatest common measure of 556, 672, and 840? *Ans.* $2^2=4$.

3. What is the greatest common measure of 110, 140, and 680? *Ans.* $2\times 5=10$.

4. What is the greatest common measure of 255, and 532? *Ans.* They have none.

5. What is the greatest common measure of 375, 408, and 922? *Ans.* They have none.

---

9. We may also find the greatest common measure of two numbers by the following

RULE.

Divide the greater by the less, then divide the divisor by the remainder, and thus continue to divide the preceding divisor by the last remainder, until there is no remainder. The last divisor will be the greatest common measure.

NOTE.—Where there is no common measure, the last divisor will be 1.

EXAMPLES.

1. What is the greatest common measure of 360 and 630?

Operation.

```
360)630(1
    360
    270)360(1
        270
         90)270(3
            270
              0
```

Hence, the greatest common measure is 90.

2. What is the greatest common measure of 922, and 408? *Ans.* 2.

3. What is the greatest common measure of 1825, and 2555? *Ans.* 365.

4. What is the greatest common measure of 124, and 682? *Ans.* 62.

5. What is the greatest common measure of 296, and 407? *Ans.* 37.

6. What is the greatest common measure of 404, and 364? *Ans.* 4.

7. What is the greatest common measure of 506, and 308? *Ans.* 22.

8. What is the greatest common measure of 212, and 416? *Ans.* 4.

9. What is the greatest common measure of 74, and 84? *Ans.* 2.

---

10. Suppose we wish to know what is the least number which will divide by 215 and 460; we proceed as follows: We decompose them into their prime factors, and thus obtain $215=5\times 43$, $460=2^2\times 5\times 23$. Hence, we see that $2^2\times 5\times 23\times 43=19780$, is the least number which can be divided by 215 and 460.

Hence, to find the least number which will divide by two or more numbers, or, as generally expressed, to find the least common multiple, we have this

RULE.

Resolve the numbers into their prime factors, (by Rule under Art. **7**)*; select all the different factors which occur, observing when the same factor has different powers, to take the highest power. The continued product of the factors thus selected will be the least common multiple.*

EXAMPLES.

1. What is the least common multiple of 12, 16, and 24?

These numbers, resolved into their prime factors, give

$$12=2^2\times 3$$
$$16=2^4$$
$$24=2^3\times 3$$

Therefore, $2^4\times 3=48$ is the least multiple required.

2. What is the least common multiple of 9, 12, 16, 20, and 35? *Ans.* $2^4\times 3^2\times 5\times 7=5040$.

3. What is the least common multiple of 7, 13, 39, and 84? *Ans.* $2^2\times 3\times 7\times 13=1092$.

4. What is the least common multiple of the nine digits? *Ans.* $2^3\times 3^2\times 5\times 7=2520$.

5. What is the least common multiple of 3, 5, 7, 12, 15, 18, and 35? *Ans.* $2^2\times 3^2\times 5\times 7=1260$.

6. What is the least common multiple of 100, 109, 463, and 900?

Ans. $2^2\times 3^2\times 5^2\times 109\times 463=45420300$.

7. What is the least common multiple of 365, 910, 2217, and 2424?

Ans. $2^3\times 3\times 5\times 7\times 13\times 73\times 101\times 739=59499225240$.

---

11. We may also find the least common multiple of two or more numbers by the following

RULE.*

Write the numbers in a horizontal line; divide them by any prime number which will divide two or more of them; place the quotients with the undivided terms for

---

* This rule is usually given as follows: "Write down the numbers in a line, and divide them by any number that will measure two or more of them, and write the quotients and undivided numbers in a line beneath. Divide this line as before, and so on, until

a second horizontal line; proceed with this second line as with the first, and so continue, until there are no two terms which can be divided. The continued product of the divisors and the numbers in the last horizontal line will be the least common multiple.

EXAMPLES.

1. What is the least common multiple of 28, 35, 42, 77, and 70?

Operation.

7	28,	35,	42,	77,	70
5	4,	5,	6,	11,	10
2	4,	1,	6,	11,	2
	2,	1,	3,	11,	1

Hence, $7\times5\times2\times2\times3\times11=4620$, is the multiple sought.

---

there are no two numbers that can be measured by the same divisor; then the continual product of all the divisors and numbers in the last line will be the least common multiple required."

The above we have copied from Mr. Adams' Arithmetic. Nearly all our Arithmetics give, in substance, the same rule. We will now show, by an example, that this rule may give very different results, depending upon the divisors used, and of course the rule is in fault.

EXAMPLE.

What is the least common multiple of 12, 16, and 24?

We will work this example in three ways, as follows:

First Operation.

12	12,	16,	24
2	1,	16,	2
	1,	8,	1

$12\times2\times8=192$.

Second Operation.

8	12,	16,	24
3	12,	2,	3
2	4,	2,	1,
	2,	1,	1

$8\times3\times2\times2=96$.

Third Operation.

4	12,	16,	24
3	3,	4,	6
2	1,	4,	2
	1,	2,	1

$4\times3\times2\times2=48$.

These operations, which are wrought strictly by this rule, give 192, 96, and 48, for the least multiple of 12, 16, and 24. Hence, the rule is wrong, and cannot be depended upon. The least common multiple of 12, 16, and 24, is 48, as may be found by either of our rules.

In the preceding example, we find, by inspection, that all the given numbers, 28, 35, 42, 77, 70, are divisible by 7, giving, for the second horizontal line, the numbers 4, 5, 6, 11, 10. Now 7 times the least multiple of 4, 5, 6, 11, 10, is the least multiple of 28, 35, 42, 77, 70, since the latter numbers are respectively 7 times the former. Again, of the numbers 4, 5, 6, 11, 10, we find that 5 and 10 are divisible by 5. Dividing, we find for the third horizontal line, the numbers 4, 1, 6, 11, 2; now, as before, 5 times the least multiple of 4, 1, 6, 11, and 2, is the least multiple of the numbers of the second line. Again, of the numbers 4, 1, 6, 11, 2, we find that 4, 6, and 2, are divisible by 2. Dividing, we obtain for the fourth horizontal line, 2, 1, 3, 11, 1; and, as before, twice the least multiple of the last numbers is the least multiple of 4, 1, 6, 11, 2, which, multiplied by 5, gives the least multiple of 4, 5, 6, 11, 10; and this result being multiplied by 7, gives the least multiple of the numbers sought. When the division is continued until there are no two terms which can be divided, the continued product of the numbers constituting the last horizontal line is the least multiple. Hence the correctness of the rule.

2. What is the least common multiple of 46, 92, 374, and 23? *Ans.* $2^2 \times 23 \times 187 = 17204$.

3. What is the least common multiple of 5, 15, 36 and 72? *Ans.* $2^3 \times 3^2 \times 5 = 360$.

4. What is the least common multiple of 11, 77, 88, and 92? *Ans.* $2^3 \times 7 \times 11 \times 23 = 14168$.

5. What is the least common multiple of 14, 51, 102, and 500? *Ans.* $2 \times 3 \times 7 \times 17 \times 250 = 178500$.

12. Suppose we wish to find all the divisors of 36, we proceed as follows: We resolve 36 into its prime factors, and thus obtain $36=2^2\times3^2$.

Now it is obvious that any combination of 2 and 3, which does not make use of these factors in a higher power than they occur in $2^2\times3^2$, must be a divisor of 36. All such combinations can be found by multiplying $1+2+4$ by $1+3+9$. Performing this multiplication, we obtain

$$\begin{array}{l} 1+2+4 \\ \underline{1+3+9} \\ 1+2+4+3+6+12+9+18+36. \end{array}$$

Therefore, the divisors of 36 are 1, 2, 3, 4, 6, 9, 12, 18 and 36.

Hence, to find all the divisors of any number, we have this

RULE.

Resolve the number into its prime factors; form as many series of terms as there are prime factors, by making 1 *the first term of any one of the series, the first power of one of the prime factors for the second term, the second power of this factor for the third term, and so on, until we reach a power as high as occurred in the decomposition. Then multiply these series together,* (*by Rule under Art.* **4**,) *and the partial products thus obtained will be the divisor sought.*

EXAMPLES.

1. What are the divisors of 48?

Here we find $48=2^4\times3$. Therefore, our series of terms will be $1+2+4+8+16$ and $1+3$; multiplying these together, (by rule under Art. **4**,) we get

$$1+2+4+8+16$$
$$1+3$$
$$\overline{1+2+4+8+16+3+6+12+24+48.}$$

Therefore, the divisors of 48 are 1, 2, 3, 4, 6, 8, 12, 16, 24, and 48.

2. What are the divisors of 360?

Ans. 1, 2, 3, 4, 5, 6, 8, 9, 10, 12, 15, 18, 20, 24, 30, 36, 40, 45, 60, 72, 90, 120, 180, 360.

3. What are the divisors of 100?

Ans. 1, 2, 4, 5, 10, 20, 25, 50, 100.

4. What are the divisors of 810?

Ans. 1, 2, 3, 5, 6, 9, 10, 15, 18, 27, 30, 45, 54, 81, 90, 135, 162, 270, 405, 810.

5. What are the divisors of 920?

Ans. 1, 2, 4, 5, 8, 10, 20, 23, 40, 46, 92, 115, 184, 230, 460, 920.

6. What are the divisors of 840?

Ans. 1, 2, 3, 4, 5, 6, 7, 8, 10, 12, 14, 15, 20, 21, 24, 28, 30, 35, 40, 42, 56, 60, 70, 84, 105, 120, 140, 168, 210, 280, 420, 840.

7. What are the divisors of 1000?

Ans. 1, 2, 4, 5, 8, 10, 20, 25, 40, 50, 100, 125, 200, 250, 500, 1000.

NOTE.—We may observe that all the divisors of these respective divisors are included in the same series, since they must evidently be divisors of the original number.

---

13. Since the series of terms which we multiplied together by the last rule, to obtain the divisors of any number commenced with 1, it follows that the number of terms in each series will be one more than the units in the exponent of the factor used.

Hence, to find the number of divisors of any number without exhibiting them, we have this

RULE.

Resolve the number into its prime factors; increase each exponent by a unit, and then take their continued product, and it will express the number of divisors.

EXAMPLES.

1. How many divisors has 4320?

$4320=2^5\times3^3\times5$. In this case, the exponents are 5, 3, and 1, each of which being increased by one, we obtain 6, 4, and 2, the continued product of which is $6\times4\times2=48$, the number of divisors sought.

2. How many divisors has 300? *Ans.* 18.
3. How many divisors has 3500? *Ans.* 24.
4. How many divisors has 162000? *Ans.* 100.
5. How many divisors has 824? *Ans.* 8.
6. How many divisors has 1172? *Ans.* 6.
7. How many divisors has 6336? *Ans.* 42.
8. How many divisors has 75600? *Ans.* 120.

CHAPTER II.

FRACTIONS.

14. A FRACTION is an expression representing a part of a unit.

VULGAR FRACTIONS.

15. A VULGAR FRACTION consists of two numbers, the one placed above the other, as in division.

The number above the line is called the *numerator;* the number below the line is called the *denominator*.

Thus, $\frac{5}{8}$ is a vulgar fraction, whose numerator is 5, and whose denominator is 8; it is read *five-eighths*.

The denominator shows how many parts the unit is divided into, and the numerator shows how many of these parts are used.

Thus, $\frac{5}{8}$ denotes that the unit is divided into 8 equal parts, and that 5 of these parts are used.

When the numerator is equal to the denominator, the fraction is equivalent to a unit. Thus, $\frac{6}{6}$, $\frac{11}{11}$, $\frac{5}{5}$, and $\frac{31}{31}$, are each equivalent to 1.

When the numerator is less than the denominator, the value of the fraction is less than a unit; it is then called a *proper fraction*.

Thus, $\frac{5}{8}$, $\frac{3}{7}$, $\frac{40}{41}$, and $\frac{20}{95}$, are each proper fractions.

When the numerator is larger than the denominator, its value is then more than a unit; it is, therefore, called an *improper fraction*.

Thus, $\frac{8}{3}$, $\frac{7}{2}$, $\frac{13}{7}$, and $\frac{40}{29}$, are each improper fractions.

A fraction of a fraction, connected by the word *of*, is called a *compound fraction*.

Thus, $\frac{2}{3}$ of $\frac{5}{8}$ of $\frac{2}{11}$ of $\frac{7}{6}$, $\frac{1}{2}$ of $\frac{2}{3}$ of $\frac{6}{7}$, and $\frac{3}{11}$ of $\frac{11}{13}$ of $\frac{2}{3}$, are compound fractions.

A fraction is said to be *inverted*, when the numerator and the denominator change places.

Thus, the fractions $\frac{16}{17}$, $\frac{11}{5}$, and $\frac{5}{8}$, when inverted, become $\frac{17}{16}$, $\frac{5}{11}$, and $\frac{8}{5}$.

Any integer may take the form of an improper fraction, by writing a unit for its denominator.

Thus, 6, 5, 3, and 11, are the same as the improper fractions $\frac{6}{1}$, $\frac{5}{1}$, $\frac{3}{1}$, and $\frac{11}{1}$.

A number consisting of an integer and a fraction, is called a *mixed number*.

Thus, $4\frac{1}{2}$, $5\frac{7}{9}$, $6\frac{3}{8}$, and $13\frac{11}{17}$, are mixed numbers. They may also be written $4+\frac{1}{2}$, $5+\frac{7}{9}$, $6+\frac{3}{8}$, and $13+\frac{11}{17}$.

REDUCTION OF FRACTIONS.

16. SINCE the value of a fraction is the quotient arising from dividing the numerator by its denominator, we may infer the following

Propositions.

I. That, multiplying the numerator of a fraction by any number, is the same as multiplying the value of the fraction by the same number.

II. That, multiplying the denominator of a fraction by any number, is the same as dividing the value of the fraction by the same number.

III. That, multiplying both numerator and denominator by the same number, does not alter the value of the fraction.

IV. That, dividing the numerator of a fraction by any number, is the same as dividing the value of the fraction by the same number.

V. That, dividing the denominator of a fraction by any number, is the same as multiplying the value of the fraction by the same number.

VI. That, dividing both numerator and denominator by the same number, does not alter the value of the fraction.

---

17. When the numerator and denominator of a fraction have no common measure, it is said to be in its *lowest terms.*

To reduce simple fractions to their lowest terms,* we have the following

---

* The following properties may frequently be of assistance in abbreviating vulgar fractions:

1. If any number terminate on the right with zero, or an *even* digit, the whole will be divisible by 2.

2. If any number terminate on the right with zero, or 5, the whole will be divisible by 5.

3. When the number expressed by the two right-hand figures are divisible by 4, the whole will be divisible by 4.

4. When the number expressed by the three right-hand figures are divisible by 8, the whole will be divisible by 8.

5. If the sum of the digits of any number be divisible by 3 or 9, then the whole number will be divisible by 3 or 9. This has already been shown under Art. 5.

6. When the difference between the sum of the digits occupying the odd places, counting from the right towards the left, and the sum of the digits occupying the even places is zero, or divisible by 11, then the number will be divisible by 11.

RULE.

Divide both numerator and denominator by their greatest common measure (found by one of the Rules under Art. **8**, *or* **9.**) *This division will not alter the value of the fraction.* (*Prop. VI., Art.* **16.**)

EXAMPLES.

1. Reduce $\frac{375}{425}$ to its lowest terms.

In this example, we find the greatest common measure of 375 and 425 to be 25.

Dividing both numerator and denominator by 25, we find $\frac{375}{425}=\frac{15}{17}$.

2. Reduce $\frac{1049}{8392}$ to its lowest terms. *Ans.* $\frac{1}{8}$.
3. Reduce $\frac{172}{1118}$ to its lowest terms. *Ans.* $\frac{2}{13}$.
4. Reduce $\frac{1740}{2900}$ to its lowest terms. *Ans.* $\frac{3}{5}$.
5. Reduce $\frac{372546}{786213}$ to its lowest terms. *Ans.* $\frac{13798}{29119}$.
6. Reduce $\frac{90062}{37956}$ to its lowest terms. *Ans.* $\frac{45031}{18978}$.
7. Reduce $\frac{432}{3080}$ to its lowest terms. *Ans.* $\frac{54}{385}$.
8. Reduce $\frac{448}{3066}$ to its lowest terms. *Ans.* $\frac{32}{219}$.
9. Reduce $\frac{3455}{4255}$ to its lowest terms. *Ans.* $\frac{691}{851}$.

Were we to resolve the numerator and denominator into their prime factors, we should then at once discover the factors common to the numerator and denominator, and could, therefore, strike them out, and then the fraction would be in its lowest terms.

For example, let it be required to reduce the fraction $\frac{728}{812}$. Resolving the numerator and denominator into their prime factors, the fraction will become $\frac{728}{812}=$ $\frac{2^3\times7\times13}{2^2\times7\times29}$. Here we discover that the factors $2^2\times7$ are common to both numerator and denominator. Striking them out, we obtain

$$\frac{728}{812}=\frac{2^3\times7\times13}{2^2\times7\times29}=\frac{2\times13}{29}=\frac{26}{29}.$$

In a similar way, we find

$$\frac{210}{495}=\frac{2\times3\times5\times7}{3^2\times5\times11}=\frac{2\times7}{3\times11}=\frac{14}{33}.$$

$$\frac{735}{2695}=\frac{3\times5\times7^2}{5\times7^2\times11}=\frac{3}{11}.$$

It is obvious that this method may be used in all cases.

Whenever we discover, by inspection, any number which will divide both numerator and denominator without a remainder, we may use it as a divisor, before resorting to either of the above methods.

---

18. To reduce an improper fraction to a mixed number, we have this

RULE.

Divide the numerator by the denominator; the quotient will be the integral part of the mixed number. The remainder, placed over the denominator of the improper fraction, will form the fractional part.

The correctness of the above rule is obvious from considering that the value of a fraction is the quotient arising from dividing the numerator by the denominator.

EXAMPLES.

1. Reduce $\frac{17}{4}$ to a mixed number.

Dividing 17 by 4, we obtain the quotient 4, with the remainder 1; $\therefore$ the mixed number equivalent to $\frac{17}{4}$, is $4\frac{1}{4}$, or $4+\frac{1}{4}$.

2. What mixed number is equivalent to $\frac{131}{7}$?

Ans. $18\frac{5}{7}$.

3. What mixed number is equivalent to $\frac{12345}{11}$?

Ans. $1122\frac{3}{11}$.

4. What mixed number is equivalent to $\frac{29817}{365}$?

Ans. $81\frac{252}{365}$.

5. What mixed number is equivalent to $\frac{70063}{6092}$?

Ans. $11\frac{3051}{6092}$.

6. What mixed number is equivalent to $\frac{3789}{292}$?

Ans. $12\frac{285}{292}$.

7. What mixed number is equivalent to $\frac{2369}{37}$?

Ans. $64\frac{1}{37}$.

8. What mixed number is equivalent to $\frac{48597354}{37538}$?

Ans. $1294\frac{23182}{37538}=1294\frac{11591}{18769}$.

9. What mixed number is equivalent to $\frac{227229}{433}$?

Ans. $524\frac{337}{433}$.

10. What mixed number is equivalent to $\frac{226072}{4583}$?

Ans. $49\frac{1505}{4583}$.

11. What mixed number is equivalent to $\frac{36573}{4927}$?

Ans. $7\frac{2084}{4927}$.

---

19. To reduce a mixed number to its equivalent improper fraction, we have this

RULE.

Multiply the integral part of the mixed number by the denominator of the fractional part; to the product add the numerator of the fractional part; the sum will be the numerator of the improper fraction, under which place the denominator of the fractional part.

This rule is obviously correct, since it is the reverse

of the rule under Art. **18**, where a reverse operation was required to be performed.

EXAMPLES.

1. Reduce $13\frac{6}{7}$ to an improper fraction.

Multiplying the integer 13 by the denominator 7, we obtain 91; to which, adding the numerator 6, we get 97 for the numerator of the improper fraction; ∴ the improper fraction equivalent to $13\frac{6}{7}$ is $\frac{97}{7}$.

2. What improper fraction is equivalent to $1278\frac{1}{3}$? *Ans.* $\frac{3835}{3}$.

3. What improper fraction is equivalent to $18910\frac{4}{7}$? *Ans.* $\frac{132374}{7}$.

4. What improper fraction is equivalent to $492536\frac{11}{13}$? *Ans.* $\frac{6402979}{13}$.

5. What improper fraction is equivalent to $5\frac{2729}{38783}$? *Ans.* $\frac{196644}{38783}$.

6. What improper fraction is equivalent to $11\frac{245}{3223}$? *Ans.* $\frac{35698}{3223}$.

7. What improper fraction is equivalent to $23\frac{27}{2329}$? *Ans.* $\frac{53594}{2329}$.

8. What improper fraction is equivalent to $375\frac{6}{773}$? *Ans.* $\frac{289881}{773}$.

9. What improper fraction is equivalent to the mixed number $6833\frac{7453}{44643}$? *Ans.* $\frac{305053072}{44643}$.

10. What improper fraction is equivalent to the mixed number $11223344\frac{5}{11}$? *Ans.* $\frac{123456789}{11}$.

---

20. Reduce the compound fraction $\frac{3}{4}$ of $\frac{7}{11}$ to its equivalent simple fraction.

$\frac{1}{4}$ of $\frac{7}{11}$ can be obtained by dividing the value of the fraction $\frac{7}{11}$ by 4, which (by Prop. II., Art. **16**,) can be effected by multiplying the denominator by 4;

$$\therefore \frac{1}{4} \text{ of } \frac{7}{11} = \frac{7}{4\times 11}.$$

Again, $\frac{3}{4}$ of $\frac{7}{11}$ is obviously three times as great as $\frac{1}{4}$ of $\frac{7}{11}$; $\therefore$ to obtain $\frac{3}{4}$ of $\frac{7}{11}$, we must multiply $\frac{7}{4\times 11}$ by 3, which (by Prop. I., Art. **16**,) can be done by multiplying the numerator by 3; hence, we have $\frac{3}{4}$ of $\frac{7}{11} = \frac{3\times 7}{4\times 11} = \frac{21}{44}$.

Hence, to reduce compound fractions to their equivalent simple ones, we have this

RULE.

Consider the word OF, *which connects the fractional parts as equivalent to the sign of multiplication. Then multiply all the numerators together for a new numerator, and all the denominators together for a new denominator, always observing to reject or cancel such factors as are common to the numerators and denominators, which is the same as dividing both numerator and denominator by the same quantity, and (by Rule under Art.* **17**,) *does not change the value of the fraction.*

EXAMPLES.

1. Reduce $\frac{1}{2}$ of $\frac{3}{4}$ of $\frac{8}{15}$ of $\frac{5}{12}$ to its equivalent simple fraction.

Substituting the sign of multiplication for the word *of*, we get $\frac{1}{2}\times\frac{3}{4}\times\frac{8}{15}\times\frac{5}{12}$. First canceling the 8 of the numerator against the 2 and 4 of the denominator, by drawing a line across them, we get $\frac{1}{\cancel{2}}\times\frac{3}{\cancel{4}}\times\frac{\cancel{3}}{5}\times\frac{5}{12}$.

Again, canceling the 3 and 5 of the numerator against the 15 of the denominator, we finally obtain

$$\frac{1}{\cancel{2}}\times\frac{\cancel{3}}{\cancel{4}}\times\frac{\cancel{3}}{\cancel{15}}\times\frac{\cancel{5}}{12}=\frac{1}{12}.$$

2. Reduce $\frac{3}{7}$ of $\frac{14}{35}$ of $\frac{7}{8}$ of $\frac{4}{9}$ of $\frac{5}{11}$ to its simplest form.

First, canceling the 7 and 5 of the numerator against the 35 of the denominator, we get $\frac{3}{7}\times\frac{14}{\cancel{35}}\times\frac{\cancel{7}}{8}\times\frac{4}{9}\times\frac{\cancel{5}}{11}$.

Again, canceling the 7 of the denominator against a part of the 14 of the numerator, and the 3 of the numerator against a part of the 9 of the denominator, we obtain

$$\frac{\cancel{3}}{\cancel{7}}\times\frac{\overset{2}{\cancel{14}}}{\cancel{35}}\times\frac{\cancel{7}}{8}\times\frac{4}{\underset{3}{\cancel{9}}}\times\frac{\cancel{5}}{11}.$$

Finally, canceling the 2 and 4 of the numerator against 8 of the denominator, we get

$$\frac{\cancel{3}}{\cancel{7}}\times\frac{\overset{\cancel{2}}{\cancel{14}}}{\cancel{35}}\times\frac{\cancel{7}}{\cancel{8}}\times\frac{\cancel{4}}{\underset{3}{\cancel{9}}}\times\frac{\cancel{5}}{11}=\frac{1}{33}.$$

NOTE.—We have written our fractions several times, in order the more clearly to exhibit the process of canceling. But in practice, it will not be necessary to write the fraction more than once. It will make no difference which of the factors are first canceled. When all the common factors have, in this way, been stricken out, the fraction will then appear in its lowest terms.

The student will find it to his interest to perform many examples of this kind, as this principle of canceling will be extensively employed in the succeeding parts of this work.

3. Reduce $\frac{3}{11}$ of $\frac{9}{13}$ of $\frac{26}{49}$ of $\frac{33}{36}$ $\frac{3}{10}$ to its simplest form. *Ans.* $\frac{27}{980}$.

4. Reduce $\frac{1}{7}$ of $\frac{7}{8}$ of $\frac{16}{13}$ of $\frac{26}{4}$ of $\frac{3}{4}$ of $\frac{4}{11}$ of $\frac{5}{1}$ to its simplest form. *Ans.* $\frac{15}{11}$.

5. Reduce $\frac{3}{24}$ of $\frac{2}{9}$ of $\frac{18}{7}$ of $\frac{14}{29}$ of $\frac{4}{18}$ to its simplest form. *Ans.* $\frac{2}{261}$.

6. Reduce $\frac{11}{23}$ of $\frac{39}{70}$ of $\frac{7}{9}$ of $\frac{88}{93}$ to its simplest form. *Ans.* $\frac{6292}{32085}$.

7. Reduce $\frac{1}{2}$ of $\frac{2}{6}$ of $\frac{3}{8}$ of $\frac{5}{22}$ of $\frac{11}{23}$ to its simplest form. *Ans.* $\frac{5}{736}$.

8. Reduce $\frac{1}{2}$ of $\frac{2}{3}$ of $\frac{3}{4}$ of $\frac{4}{5}$ of $\frac{5}{6}$ of $\frac{6}{7}$ of $\frac{7}{8}$ of $\frac{8}{9}$ to its simplest form. *Ans.* $\frac{1}{9}$.

9. Reduce $\frac{7}{23}$ of $\frac{11}{8}$ of $\frac{26}{55}$ of $\frac{10}{22}$ to its simplest form. *Ans.* $\frac{91}{1012}$.

10. Reduce $\frac{3}{7}$ of $\frac{14}{25}$ of $\frac{10}{27}$ of $\frac{34}{35}$ to its simplest form. *Ans.* $\frac{136}{1575}$.

---

21. To reduce fractions to a common denominator, we have this

RULE.

Reduce mixed numbers to improper fractions—compound fractions to their simplest form. Then multiply each numerator by all the denominators, except its own, for a new numerator, and all the denominators together for a common denominator.

It is obvious that this process will give the same denominator to each fraction, viz: the product of all the denominators.

It is also obvious that the values of the fractions will not be changed, since both numerator and denominator are multiplied by the same quantity, viz: the product of all the denominators except its own.

EXAMPLES.

1. Reduce $\frac{1}{2}$, $\frac{5}{3}$ of $\frac{2}{5}$, $\frac{3}{11}$, and $\frac{7}{9}$ of $\frac{2}{7}$, to equivalent fractions having a common denominator.

These fractions, when reduced to their simplest form, are $\frac{1}{2}$, $\frac{2}{3}$, $\frac{3}{11}$, and $\frac{2}{9}$.

The new numerator of the first fraction is $1\times3\times11\times9=297$.

The new numerator of the second fraction is $2\times2\times11\times9=396$.

The new numerator of the third fraction is $3\times2\times3\times9=162$.

The new numerator of the fourth fraction is $2\times2\times3\times11=122$.

The common denominator is $2\times3\times11\times9=594$.

Therefore, the fractions, when reduced to a common denominator, are $\frac{297}{594}$, $\frac{396}{594}$, $\frac{162}{594}$, and $\frac{132}{594}$.

2. Reduce $\frac{3}{8}$ of $\frac{7}{3}$, $\frac{6}{11}$ of $\frac{11}{7}$, and $\frac{40}{41}$, to equivalent fractions having a common denominator.

Ans. $\frac{2009}{2296}$, $\frac{1968}{2296}$, and $\frac{2240}{2296}$.

3. Reduce $\frac{7}{11}$, $\frac{13}{37}$ of $\frac{37}{17}$, and $\frac{41}{43}$, to equivalent frac tions having a common denominator.

Ans. $\frac{5117}{8041}$, $\frac{6149}{8041}$, and $\frac{7667}{8041}$.

4. Reduce $\frac{23}{29}$, $\frac{31}{47}$, $\frac{41}{43}$, and $\frac{47}{53}$ to equivalent fractions having a common denominator.

Ans. $\frac{2463599}{3106277}$, $\frac{2048821}{3106277}$, $\frac{2961799}{3106277}$, $\frac{2754623}{3106277}$.

5. Reduce $\frac{73}{79}$, $\frac{83}{89}$, and $\frac{97}{101}$, to fractions having a common denominator. *Ans.* $\frac{656197}{710131}$, $\frac{662257}{710131}$, $\frac{682007}{710131}$.

6. Reduce $\frac{173}{977}$, and $\frac{1109}{1123}$, to fractions having a common denominator. *Ans.* $\frac{194279}{1097171}$, $\frac{1083493}{1097171}$.

22. To reduce fractions to their least common denominators, we have this

RULE.

Reduce the fractions to their simplest form. Then find the least common multiple of their denominators, (by Rule under Art. **10**, *or Rule under Art.* **11**,) *which will be their least common denominator. Divide this common denominator by the respective denominators of the given fractions; multiply the quotients by their respective numerators, and the products will be the new numerator.*

The correctness of the above rule may be shown in the same way as was that of the preceding rule.

EXAMPLES.

1. Reduce $\frac{1}{2}$ of $\frac{3}{7}$ of $\frac{7}{12}$, $\frac{3}{20}$, and $\frac{7}{15}$, to equivalent fractions having the least common denominator.

These fractions, when reduced to their simplest form, become $\frac{1}{8}$, $\frac{3}{20}$, and $\frac{7}{15}$. The least common multiple of the denominators 8, 20, and 15, is 120=common denominator.

New numerator of the first fraction is $\frac{120}{8}\times1=15$.
" " of the second fraction is $\frac{120}{20}\times3=18$.
" " of the third fraction is $\frac{120}{15}\times7=56$.

Hence, the fractions, when reduced to their least common denominator, become $\frac{15}{120}$, $\frac{18}{120}$, and $\frac{56}{120}$.

2. Reduce $\frac{7}{8}$ of $\frac{4}{5}$, $4\frac{1}{2}$, and $\frac{3}{20}$, to equivalent fractions having the least common denominator.

Ans. $\frac{14}{20}$, $\frac{90}{20}$, and $\frac{3}{20}$.

3. Reduce $\frac{7}{8}$ of $\frac{3}{14}$ of $\frac{16}{13}$ of $\frac{26}{10}$, $\frac{11}{20}$, and $7\frac{1}{2}$, to fractions having the least common denominator.

Ans. $\frac{12}{20}$, $\frac{11}{20}$, and $\frac{150}{20}$.

4. Reduce $\frac{11}{13}$ of $\frac{39}{44}$ of $\frac{4}{18}$ of $\frac{12}{7}$, $6\frac{1}{7}$, and $\frac{8}{28}$, to fractions having the least common denominator.

Ans. $\frac{2}{7}$, $\frac{43}{7}$, and $\frac{2}{7}$.

5. Reduce $\frac{3}{7}$ of $\frac{22}{28}$ of $\frac{9}{22}$ of $\frac{55}{49}$, $\frac{6}{22}$ of $\frac{3}{32}$, and $\frac{3}{33}$ of $\frac{22}{36}$, to fractions having the least common denominator.

Ans. $\frac{1176120}{7606368}$, $\frac{1944481}{7606368}$, $\frac{4225576}{7606368}$.

6. Reduce $\frac{11}{7}$, $11\frac{5}{7}$, and $7\frac{5}{33}$, to fractions having the least common denominator. *Ans.* $\frac{363}{231}$, $\frac{2706}{231}$, $\frac{1652}{231}$.

7. Reduce $\frac{1}{2}$ of $\frac{4}{7}$ of $\frac{24}{30}$ of $\frac{30}{48}$, $3\frac{2}{28}$, and $10\frac{3}{56}$, to fractions having the least common denominator.

Ans. $\frac{8}{56}$, $\frac{172}{56}$, $\frac{563}{56}$.

8. Reduce $\frac{17}{38}$, $\frac{13}{43}$ of $\frac{86}{95}$, and $\frac{37}{47}$, to fractions having the least common denominator. *Ans.* $\frac{3995}{8930}$, $\frac{2444}{8930}$, $\frac{7030}{8930}$.

9. Reduce $3\frac{1}{7}$, $5\frac{1}{5}$, $5\frac{1}{4}$, and $11\frac{9}{33}$, to fractions having the least common denominator.

Ans. $\frac{4840}{1540}$, $\frac{8008}{1540}$, $\frac{8085}{1540}$, $\frac{17360}{1540}$.

ADDITION OF FRACTIONS.

23. Suppose we wish to add $\frac{3}{7}$ and $\frac{4}{5}$. We know that so long as these fractions are of different denominators they cannot be added; we will, therefore, reduce them to a common denominator; we thus obtain $\frac{3}{7}=\frac{15}{35}$, $\frac{4}{5}=\frac{28}{35}$. Now, taking their sum, we get $\frac{3}{7}+\frac{4}{5}=\frac{15}{35}+\frac{28}{35}=\frac{15+28}{35}=\frac{43}{35}=1\frac{8}{35}$.

Hence, to add fractions, we have this

RULE.

Reduce the fractions to a common denominator, and take the sum of the numerators, under which place the common denominator, and it will give the sum required.

EXAMPLES.

1. Add the fractions $\frac{1}{2}$ of $\frac{5}{7}$, $\frac{3}{11}$ of $\frac{11}{6}$, and $5\frac{1}{7}$.

These fractions, reduced to their least common denominator, are $\frac{5}{14}$, $\frac{7}{14}$, and $\frac{72}{14}$; and their sum is

$$\frac{5+7+72}{14}=\frac{84}{14}=6.$$

2. Add the fractions $\frac{3}{10}$, $\frac{7}{5}$, $\frac{8}{15}$, and $\frac{1}{30}$.

Ans. $\frac{68}{30}=2\frac{4}{15}$.

3. Add the fractions $\frac{5}{7}$ of $\frac{7}{8}$, $\frac{3}{16}$, and $4\frac{1}{2}$.

Ans. $\frac{85}{16}=5\frac{5}{16}$.

4. Add the fractions $\frac{3}{4}$, $\frac{4}{5}$, $\frac{13}{17}$, and $\frac{2}{11}$.

Ans. $\frac{9337}{3740}=2\frac{1857}{3740}$.

5. What is the sum of $\frac{7}{23}$, $\frac{13}{23}$, and $\frac{43}{92}$?

Ans. $\frac{123}{92}=1\frac{31}{92}$.

6. What is the sum of $\frac{240}{329}$, $\frac{722}{689}$, and $\frac{1}{3}$?

Ans. $\frac{1435375}{680043}=2\frac{75289}{680043}$.

7. What is the sum of $\frac{49}{84}$, $\frac{23}{22}$, and $\frac{7}{9}$?

Ans. $\frac{953}{396}=2\frac{161}{396}$.

8. What is the sum of $\frac{1}{2}$ of $4\frac{1}{5}$, $\frac{3}{22}$ of $6\frac{1}{7}$, and $\frac{5}{8}$?

Ans. $\frac{10973}{3080}=3\frac{1733}{3080}$.

9. What is the sum of $\frac{1}{2}$, $\frac{1}{3}$, $\frac{1}{4}$, $\frac{1}{5}$, $\frac{2}{6}$, and $\frac{1}{20}$?

Ans. $\frac{100}{60}=1\frac{2}{3}$.

10. What is the sum of $\frac{1}{2}$, $\frac{2}{3}$, $\frac{3}{4}$, $\frac{4}{5}$, $\frac{5}{6}$, and $\frac{9}{20}$?

Ans. 4.

11. What is the sum of $1\frac{1}{2}$, $2\frac{1}{3}$, $3\frac{1}{4}$, $4\frac{1}{5}$, and $5\frac{1}{6}$?

Ans. $16\frac{9}{20}$.

SUBTRACTION OF FRACTIONS.

24. To subtract one fraction from another, we have this

RULE.

Reduce the fractions to a common denominator, and subtract the numerator of the subtrahend from that of the minuend; place the common denominator under the difference.

EXAMPLES.

1. From $\frac{3}{11}$ subtract $\frac{3}{31}$.

These fractions, when reduced to their least common denominator, become $\frac{93}{341}$, $\frac{88}{341}$. Therefore, $\frac{3}{11}-\frac{8}{31}=\frac{93}{341}-\frac{88}{341}=\frac{5}{341}$.

2. From $\frac{17}{43}$ subtract $\frac{4}{29}$ *Ans.* $\frac{321}{1247}$.
3. From $2\frac{7}{8}$ subtract $\frac{19}{16}$ *Ans.* $\frac{27}{16}=1\frac{11}{16}$.
4. From $\frac{3}{4}$ of $\frac{4}{12}$ of $\frac{6}{7}$ subtract $\frac{1}{2}$ of $\frac{3}{19}$. *Ans.* $\frac{72}{532}$.
5. From $\frac{3}{17}$ subtract $\frac{1}{43}$. *Ans.* $\frac{112}{731}$.
6. From $\frac{49}{103}$ subtract $\frac{71}{1063}$. *Ans.* $\frac{44774}{109489}$.
7. From $\frac{13}{73}$ subtract $\frac{4}{62}$. *Ans.* $\frac{257}{2263}$.
8. From $3\frac{1}{3}$ subtract $2\frac{3}{5}$. *Ans.* $\frac{11}{15}$.
9. From $\frac{1}{7}$ of $4\frac{1}{9}$ subtract $\frac{3}{8}$. *Ans.* $\frac{107}{504}$.
10. From $\frac{22}{36}$ of $\frac{2}{49}$ subtract $\frac{3}{34}$ of $\frac{2}{8}$. *Ans.* $\frac{173}{59976}$.
11. From $\frac{6}{33}$ of $\frac{2}{3}$ subtract $\frac{2}{29}$. *Ans.* $\frac{50}{957}$.
12. From $\frac{67}{47}$ subtract $\frac{22}{37}$. *Ans.* $\frac{1445}{1739}$.

MULTIPLICATION OF FRACTIONS.

25. LET it be required to multiply $\frac{3}{4}$ by $\frac{5}{7}$.

We have seen (under Art. **20**,) that $\frac{3}{4}$ multiplied by $\frac{5}{7}$ is the same as $\frac{3}{4}$ of $\frac{5}{7}$. Therefore, we must use the same rule for multiplying fractions as for reducing compound fractions.

Hence, to multiply together fractions, we have this

RULE.

Multiply all the numerators together for a new numerator, and all the denominators together for a new denominator, always observing to reject or cancel such factors as are common to both numerators and denominators.

EXAMPLES.

1. Multiply together the fractions $\frac{3}{11}$, $\frac{22}{21}$, $\frac{7}{9}$, and $\frac{1}{3}$.

Expressing the multiplication, we obtain $\frac{3}{11}\times\frac{22}{21}\times\frac{7}{9}\times\frac{1}{3}$.

Canceling the 3 and 7 of the numerators, against 21 of the denominators, also the 11 of the denominators against a part of the 22 of the numerators, we get

$$\frac{\cancel{3}}{\cancel{11}}\times\frac{\overset{2}{\cancel{22}}}{\cancel{21}}\times\frac{\cancel{7}}{9}\times\frac{1}{3}=\frac{2}{9\times3}=\frac{2}{27}.$$

2. Multiply together the fractions $\frac{7}{11}$, $\frac{21}{35}$, $\frac{55}{42}$, and $\frac{4}{9}$.

Indicating the multiplication, we get $\frac{7}{11}\times\frac{21}{35}\times\frac{55}{42}\times\frac{4}{9}$. Canceling the 11 of the denominators, against a part of the 55 of the numerators, also the 7 of the numerators, against a part of the 35 of the denominators, we obtain

$$\frac{\cancel{7}}{\cancel{11}}\times\frac{21}{\underset{5}{\cancel{35}}}\times\frac{\overset{5}{\cancel{55}}}{42}\times\frac{4}{9}.$$

Again, canceling the 5, which is common to both numerators and denominators, also the factor 7, which is common to 21 of the numerators, and to 42 of the denominators, we get

$$\frac{\cancel{7}}{\cancel{11}}\times\frac{\overset{3}{\cancel{21}}}{\underset{\cancel{5}}{\cancel{35}}}\times\frac{\overset{\cancel{5}}{\cancel{55}}}{\underset{6}{\cancel{42}}}\times\frac{4}{9}.$$

Finally, canceling the 3 of the numerators, against a part of the 9 of the denominators, and the factor 2, which is common to the 4 of the numerators, and to the 6 of the denominators, we obtain

$$\frac{\cancel{7}}{\cancel{11}}\times\frac{\overset{\cancel{3}}{\cancel{21}}}{\underset{\cancel{5}}{\cancel{35}}}\times\frac{\overset{\cancel{5}}{\cancel{55}}}{\underset{\underset{3}{\cancel{6}}}{\cancel{42}}}\times\frac{\overset{2}{\cancel{4}}}{\underset{3}{\cancel{9}}}=\frac{2}{3\times3}=\frac{2}{9}.$$

NOTE.—A little practice will enable the student to perform these operations of canceling with great ease and rapidity. And since, as was remarked under Art. 20, it is immaterial which factors are first canceled, the simplicity of the work must depend much upon his skill or ingenuity.

3. Multiply together the fractions $\frac{3}{40}$, $\frac{7}{18}$, and $\frac{10}{14}$.
Ans. $\frac{1}{48}$.

4. Multiply together the fractions $\frac{17}{39}$, $\frac{14}{34}$, $\frac{26}{28}$, $\frac{1}{11}$.
Ans. $\frac{1}{66}$.

5. Multiply together the fractions $\frac{3}{7}$, $3\frac{1}{3}$, $\frac{1}{8}$ of $\frac{16}{33}$, and $\frac{1}{11}$.
Ans. $\frac{20}{2541}$.

6. Multiply together the fractions $\frac{2}{3}$, $\frac{4}{5}$, $\frac{6}{7}$, and $\frac{2}{11}$.
Ans. $\frac{32}{385}$.

7. Multiply together $\frac{3}{7}$, $\frac{4}{9}$, $\frac{18}{25}$, and $\frac{5}{6}$. *Ans.* $\frac{4}{35}$.

8. Multiply together $\frac{3}{10}$, $\frac{10}{11}$, $\frac{22}{27}$, and $\frac{1}{2}$. *Ans.* $\frac{1}{9}$.

9. Multiply together $\frac{7}{8}$, $\frac{3}{7}$, $\frac{8}{9}$, $\frac{1}{2}$, $\frac{3}{4}$, and $\frac{1}{5}$. *Ans.* $\frac{1}{40}$.

10. Multiply together $\frac{2}{3}$, $\frac{3}{5}$, $\frac{4}{9}$, $\frac{9}{14}$, $\frac{3}{8}$, and $\frac{6}{5}$.
Ans. $\frac{9}{175}$.

DIVISION OF FRACTIONS.

26. Let it be required to divide $\frac{4}{7}$ by $\frac{5}{8}$.

We know that $\frac{4}{7}$ can be divided by 5, by multiplying the denominator by 5, (see Prop. II., Art. **16**,) which gives $\frac{4}{7\times 5}$.

Now, since $\frac{5}{8}$ is but one-eighth of 5, it follows that $\frac{4}{7}$, divided by $\frac{5}{8}$, must be eight times as great as $\frac{4}{7}$ divided by 5. $\therefore$ $\frac{4}{7}$, divided by $\frac{5}{8}$, must be $\frac{4\times 8}{7\times 5}$. From this, we see that $\frac{4}{7}$ has been multiplied by $\frac{5}{8}$, when inverted.

Hence, to divide one fraction by another, we have this

RULE.

Reduce the fractions to their simplest form. Invert the divisor, and then proceed as in multiplication.

EXAMPLES.

1. Divide $\frac{43}{84}$ by $\frac{86}{21}$.

Inverting the divisor, and then multiplying, we obtain $\frac{43}{84}\times\frac{21}{86}$; which, by canceling, becomes $\frac{\not{4}\not{3}}{\underset{4}{\not{8}\not{4}}}\times\frac{\not{2}\not{1}}{\underset{2}{\not{8}\not{6}}}=\frac{1}{8}$.

2. Divide $\frac{107}{209}$ by $\frac{7}{418}$. *Ans.* $\frac{214}{7}=30\frac{4}{7}$.
3. Divide $\frac{37}{49}$ by $\frac{98}{365}$. *Ans.* $\frac{13505}{4802}=2\frac{3901}{4802}$.
4. Divide $\frac{15}{17}$ by $\frac{30}{3}$. *Ans.* $\frac{3}{34}$.
5. Divide $4\frac{1}{3}$ by $17\frac{1}{2}$. *Ans.* $\frac{26}{105}$.

6. Divide $\frac{1}{2}$ of $4\frac{1}{7}$ by $\frac{2}{3}$ of $\frac{5}{8}$. *Ans.* $\frac{174}{35}=4\frac{34}{35}$.

7. Divide $\frac{3}{7}$ of $\frac{3}{8}$ of 7 by $\frac{13}{3}$ of 6. *Ans.* $\frac{9}{208}$.

8. Divide $\frac{14}{37}$ by $\frac{28}{35}$. *Ans.* $\frac{35}{74}$.

9. Divide $\frac{1}{2}$ of $\frac{2}{3}$ of $\frac{3}{4}$ by $\frac{2}{3}$ of $\frac{4}{3}$. *Ans.* $\frac{9}{32}$.

10. Divide $\frac{1}{20}$ of $\frac{1}{22}$ of $\frac{22}{7}$ by $\frac{3}{34}$. *Ans.* $\frac{17}{210}$.

COMPLEX FRACTIONS.

27. Sometimes fractions occur, in which the numerator, or denominator, or both, are already fractional.

Thus, $\dfrac{2}{\frac{3}{7}}$, $\dfrac{\frac{2}{3}}{5}$, $\dfrac{\frac{3}{4}}{\frac{7}{8}}$, $\dfrac{\frac{1}{9}}{\frac{1}{13}}$: such fractions are called *complex fractions*.

REDUCTION OF COMPLEX FRACTIONS.

28. Since the value of a fraction is the quotient arising from dividing the numerator by the denominator, it follows that the complex fraction $\dfrac{2}{\frac{3}{7}}$ is the same as

$$2\div\tfrac{3}{7}=\tfrac{14}{3}=4\tfrac{2}{3}.\quad \text{Again, } \dfrac{\frac{3}{4}}{\frac{7}{8}}=\tfrac{3}{4}\div\tfrac{7}{8}=\tfrac{6}{7}.$$

Hence, to reduce a complex fraction to a simple one, we have this

RULE.

Divide the numerator of the complex fraction by the denominator, according to Rule under Art. **26.**

EXAMPLES.

1. Reduce $\dfrac{4\frac{1}{2}}{3\frac{1}{3}}$ to a simple fraction

Dividing $4\frac{1}{2}=\frac{9}{2}$ by $3\frac{1}{3}=\frac{10}{3}$, we get $\frac{27}{20}=1\frac{7}{20}$.

2. Reduce $\frac{\frac{3}{7}}{\frac{6}{9}}$ to a simple fraction. *Ans.* $\frac{9}{14}$

3. Reduce $\frac{3\frac{1}{7}}{\frac{4}{5}}$ to a simple fraction. *Ans.* $\frac{55}{14}=3\frac{13}{14}$

4. Reduce $\frac{6\frac{1}{7}}{3\frac{1}{4}}$ to a simple fraction.

Ans. $\frac{172}{91}=1\frac{81}{91}$.

5. Reduce $\frac{\frac{13}{4}}{4\frac{1}{2}}$ to a simple fraction. *Ans.* $\frac{13}{18}$.

6. Reduce $\frac{6\frac{9}{10}}{8}$ to a simple fraction. *Ans.* $\frac{69}{80}$.

7. Reduce $\frac{1\frac{3}{8}}{\frac{1}{2}\text{ of }\frac{1}{3}}$ to a simple fraction.

Ans. $\frac{33}{4}=8\frac{1}{4}$.

8. Reduce $\frac{\frac{3}{4}\text{ of }\frac{7}{9}}{\frac{2}{3}\text{ of }\frac{1}{5}}$ to a simple fraction. *Ans.* $\frac{35}{8}=4\frac{3}{8}$

9. Reduce $\frac{6+\frac{1}{8}}{7+\frac{1}{2}}$ to a simple fraction. *Ans.* $\frac{49}{60}$.

10. Reduce $\frac{10\frac{1}{23}}{11\frac{1}{32}}$ to a simple fraction. *Ans.* $\frac{7392}{8119}$.

REDUCTION OF FRACTIONS

TO A GIVEN DENOMINATOR.

29. SUPPOSE we wish to change the fraction $\frac{4}{5}$ to an equivalent one, having 6 for its denominator.

It is obvious that if we first multiply $\frac{4}{5}$ by 6, and then divide the product by 6, its value will not be altered. By this means, we find that $\frac{4}{5}=\frac{\frac{4}{5}\times 6}{6}=\frac{\frac{24}{5}}{6}=\frac{4\frac{4}{5}}{6}$.

Hence, to reduce a fraction to an equivalent one having a given denominator, we have this

RULE.

Multiply the fraction by the number which is to be the given denominator, (see Rule under Art. **25**,*) under which place the given denominator, and it will be the fraction required.*

EXAMPLES.

1. Reduce $\frac{3}{7}$ to an equivalent fraction having 8 for its denominator.

In this example, we first multiply $\frac{3}{7}$ by 8, which gives $\frac{24}{7}$; therefore, placing 8 under $\frac{24}{7}$, we get $\frac{\frac{24}{7}}{8}=\frac{3\frac{3}{7}}{8}$ for the fraction required.

2. Reduce $\frac{3}{11}$ to an equivalent fraction having 12 for its denominator. *Ans.* $\frac{3\frac{3}{11}}{12}$.

3. Reduce $\frac{13}{10}$ to an equivalent fraction having 7 for its denominator. *Ans.* $\frac{9\frac{1}{10}}{7}$.

4. Reduce $\frac{1}{2}$, $\frac{3\frac{1}{3}}{10}$, $\frac{1}{4}$, and $\frac{1\frac{2}{3}}{10}$ to fractions having 12 for their common denominators.

Ans. $\frac{6}{12}$, $\frac{4}{12}$, $\frac{3}{12}$, and $\frac{2}{12}$

5. Reduce $\frac{1}{9}$, $\frac{1}{11}$, $\frac{1}{13}$, and $\frac{1}{17}$, to fractions having 100 for their common denominator.

Ans. $\frac{11\frac{1}{9}}{100}$, $\frac{9\frac{1}{11}}{100}$, $\frac{7\frac{9}{13}}{100}$, and $\frac{5\frac{15}{17}}{100}$.

6. Reduce $\frac{1}{2}$, $\frac{1}{3}$, $\frac{1}{4}$, $\frac{1}{5}$, and $\frac{1}{6}$, to fractions having 30 for their common denominator.

Ans. $\frac{15}{30}$, $\frac{10}{30}$, $\frac{7\frac{1}{2}}{30}$, $\frac{6}{30}$, and $\frac{5}{30}$.

REDUCTION OF DENOMINATE FRACTIONS.

30. A *denominate fraction* is a fraction of a number of a particular denomination. Thus, $\frac{7}{8}$ of a foot, $\frac{3}{4}$ of a yard, $\frac{1}{3}$ of a dollar, and $\frac{5}{9}$ of a shilling are denominate fractions.

Reduction of denominate fractions is the changing of them from one denomination to another, without altering their values.

---

31. Suppose we wish to reduce $\frac{1}{360}$ of a pound sterling to an equivalent fraction of a farthing, we proceed as follows: Since there are 20 shillings in 1 pound, it follows that $\frac{1}{360}$ of a pound is the same as 20 times $\frac{1}{360}$ of a shilling, and this is also the same as 12 times 20 times $\frac{1}{360}$ of a penny; which, in turn, is 4 times 12 times 20 times $\frac{1}{360}$ of a farthing. Hence, $\frac{1}{360}$ of a pound sterling is equivalent to $\frac{1}{360}$ of $\frac{20}{1}$ of $\frac{12}{1}$ of $\frac{4}{1}$ of a farthing.

Again, let us reduce $\frac{3}{5}$ of a farthing to an equivalent fraction of a pound sterling. In this case, we must use the reciprocals of $\frac{20}{1}$, $\frac{12}{1}$, $\frac{4}{1}$; we thus find that $\frac{3}{5}$ of a farthing is equivalent to $\frac{3}{5}$ of $\frac{1}{4}$ of $\frac{1}{12}$, of $\frac{1}{20}$ of a pound sterling.

Hence, to reduce fractions of one denominate value to equivalent fractions of other denominate values, we have this

RULE.

I. *When the given fraction is to be reduced to a higher denomination, multiply it by a compound fraction whose terms are the reciprocals of the successive denominate values, included between the denomination of the given fraction, and the one to which it is to be reduced.*

II. *When the given fraction is to be reduced to a lower denomination, then multiply it by a compound fraction whose terms have units for their denominators, and for numerators the successive denominate values included between the denomination of the given fraction and the one to which it is to be reduced.*

EXAMPLES.

1. Reduce $\frac{3}{8}$ of an inch to the fraction of a mile.

In this example, the different denominate values between an inch and a mile are 12 inches, $16\frac{1}{2}=\frac{33}{2}$ feet, 40 rods, and 8 furlongs; ∴ our compound fraction is $\frac{1}{12}$ of $\frac{2}{33}$ of $\frac{1}{40}$ of $\frac{1}{8}$; which, multiplied by the given fraction, produces $\frac{3}{8}$ of $\frac{1}{12}$ of $\frac{2}{33}$ of $\frac{1}{40}$ of $\frac{1}{8}$; canceling the 3 and 2 of the numerators, against a part of the 12 of the denominators, we get,

$$\frac{\not{3}}{8}\times\frac{1}{\underset{2}{\not{12}}}\times\frac{\not{2}}{33}\times\frac{1}{40}\times\frac{1}{8}=\frac{1}{168960}.$$

Therefore, $\frac{3}{8}$ of an inch is equivalent to $\frac{1}{168960}$ of a mile.

2. Reduce $\frac{3}{11520}$ of a solar day to an equivalent fraction of a second.

In this example, the successive denominate values between a solar day and a second, are 24 hours, 60 minutes, and 60 seconds; therefore, our compound fraction is $\frac{24}{1}$ of $\frac{60}{1}$ of $\frac{60}{1}$; which, multiplied by the given fraction, becomes $\frac{3}{11520}$ of $\frac{24}{1}$ of $\frac{60}{1}$ of $\frac{60}{1}$; this becomes, after canceling like factors, $\frac{45}{2}$ of a second.

3. Reduce $\frac{144}{360}$ of a yard to the fraction of a mile.
Ans. $\frac{1}{4400}$.

4. Reduce $\frac{27}{84}$ of a gill to the fraction of a gallon.
Ans. $\frac{9}{896}$.

5. Reduce $\frac{330}{495}$ of a pound to the fraction of a ton.
Ans. $\frac{1}{3360}$.

6. Reduce $\frac{1}{3}$ of a mile to feet. *Ans.* 1760 feet.

7. Reduce $\frac{1}{2}$ of $\frac{1}{3}$ of $\frac{2}{7}$ of a yard to the fraction of a mile. *Ans.* $\frac{1}{36960}$.

8. Reduce $\frac{1}{6}$ of $\frac{1}{8}$ of $\frac{24}{28}$ of a gallon to the fraction of a gill. *Ans.* $\frac{4}{7}$.

9. Reduce $\frac{2}{3}$ of $\frac{4}{9}$ of a hogshead of wine to the fraction of a gill. *Ans.* $\frac{1792}{3}=597\frac{1}{3}$ gills.

10. Reduce $\frac{1}{2}$ of $\frac{3}{7}$ of $4\frac{1}{2}$ yards to the fraction of an inch. *Ans.* $\frac{243}{7}=34\frac{5}{7}$ inches.

11. Reduce $\frac{1}{7}$ of $\frac{3}{43}$ of a farthing to the fraction of a shilling. *Ans.* $\frac{1}{4816}$.

12. Reduce $\frac{7}{59}$ of an ounce to the fraction of a pound avoirdupois. *Ans.* $\frac{7}{944}$.

---

32. To find what fractional part one quantity is of another of the same kind, but of different denominations.

Suppose we wish to know what part of 1 yard 2 feet

3 inches is; we reduce 1 yard to inches, which gives 1 yard=36 inches; we also reduce 2 feet 3 inches to inches, which gives 2 feet 3 inches=27 inches. Now it is obvious that 2 feet 3 inches is the same part of one yard that 27 is of 36, which is $\frac{27}{36}=\frac{3}{4}$.

Hence, we deduce this

RULE.

Reduce the given quantities to the lowest denomination mentioned in either; then divide the number, which is to become the fractional part, by the other number.

EXAMPLES.

1. What part of £3 4 s. 1 d. is 2 s. 6 d.?

In this example, the quantities, when reduced, become £3 4 s. 1 d.=769 d.; and 2 s. 6 d.=30 d.; therefore $\frac{30}{769}$ is the fractional part which 2 s. 6 d. is of £3 4 s. 1 d.

2. What part of 3 miles 40 rods is 27 feet 9 inches?
Ans. $\frac{37}{22000}$.

3. What part of a day is 17 minutes 4 seconds?
Ans. $\frac{8}{675}$.

4. What part of \$700 is \$5.30? *Ans.* $\frac{53}{7000}$.

5. What fractional part of 2 hogsheads is 3 pints?
Ans. $\frac{1}{336}$.

6. What part of \$3 is $2\frac{1}{2}$ cents? *Ans.* $\frac{1}{120}$.

7. What part of 10 shillings 8 pence is 3 shillings 1 penny? *Ans.* $\frac{37}{128}$.

8. What part of 100 acres is 63 acres, 2 roods, and 7 rods of land? *Ans.* $\frac{10167}{16000}$.

33. To reduce a fraction of any given denomination to whole denominate numbers.

Suppose we wish to know the value of $\frac{3}{8}$ of a yard; we know that $\frac{3}{8}$ of a yard equals $\frac{3}{8}$ of $\frac{4}{1}$ of a quarter $= \frac{3}{2}$ of a quarter $= 1$ quarter $+ \frac{1}{2}$ of a quarter.

Again, $\frac{1}{2}$ of a quarter equals $\frac{1}{2}$ of $\frac{4}{1}$ of a nail $= 2$ nails. Therefore, $\frac{3}{8}$ of a yard equals 1 quarter 2 nails. Hence, we deduce this

RULE.

Multiply the numerator by the units in the next inferior denominate value, and divide the product by the denominator; multiply the remainder, if any, by the next lower denominate value, and again divide the product by the denominator; continue this process until there is no remainder, or until we reach the lowest denominate value. The successive quotients will form the successive denominate values.

EXAMPLES.

1. What is the value of $\frac{3}{15}$ of an hour?

In this example, $\frac{3}{15}$ of an hour $= \frac{3}{15}$ of $\frac{60}{1}$ of a minute $= 12$ minutes.

2. What is the value of $\frac{3}{7}$ of 1 yard?

Ans. 1 quarter, $2\frac{6}{7}$ nails.

3. What is the value of $\frac{1}{2}$ of $\frac{3}{8}$ of one mile?

Ans. 1 furlong, 20 rods.

4. What is the value of $\frac{3}{7}$ of $\frac{5}{6}$ of 1 cwt.?

Ans. 1 quarter, 12 pounds.

5. What is the value of $\frac{1}{6}$ of 14 miles, 6 furlongs?

Ans. 2 miles, 3 furlongs, 26 rods, 11 feet.

6. What is the value of $\frac{1}{3}$ of $\frac{3}{5}$ of 2 days of 24 hours each? *Ans.* 9 hours, 36 minutes.

7. What is the value of $\frac{1}{2}$ of $\frac{7}{8}$ of $\frac{3}{14}$ of an hour?
Ans. 5 minutes, $37\frac{1}{2}$ seconds.

8. What is the value of $\frac{1}{2}$ of $\frac{1}{3}$ of 9 hours and 18 minutes? *Ans.* 1 hour, 33 minutes.

CHAPTER III.

DECIMAL FRACTIONS.

34. A DECIMAL FRACTION is that particular form of a vulgar fraction whose denominator consists of a unit followed by one or more ciphers.

Thus, $\frac{6}{10}$, $\frac{37}{100}$, $\frac{4}{1000}$, and $\frac{3634}{10000}$, are decimal fractions.

In practice, the denominators of decimal fractions are not written, but they are always understood.

Thus, instead of $\frac{3}{10}$, $\frac{7}{100}$, $\frac{37}{1000}$, and $\frac{1}{10000}$, we write 0·3, 0·07, 0·037, and 0·0001.

The first figure on the right of the period, or *decimal point*, is said to be in the place of tenths, the second figure is said to be in the place of hundredths, the third in the place of thousandths, and so on, decreasing from the left towards the right, in a ten-fold ratio, the same as in whole numbers. The following table will exhibit this more clearly.

NUMERATION TABLE

OF WHOLE NUMBERS AND DECIMALS.*

&c. &c.	Tens of Billions.	Billions.	Hundreds of Millions.	Tens of Millions.	Millions.	Hundreds of Thousands.	Tens of Thousands.	Thousands.	Hundreds.	Tens.	UNITS.	Decimal Point.	Tenths.	Hundredths.	Thousandths.	Ten Thousandths.	Hundred Thousandths.	Millionths.	Ten Millionths.	Hundred Millionths.	Billionths.	Ten Billionths.	&c. &c.
	3	3	3	3	3	3	3	3	3	3	3	·	3	3	3	3	3	3	3	3	3	3	

ASCENDING. ☜ ☞ DESCENDING.

* This table is in accordance with the French method of numbering, where each period of three figures changes its denominate value.

EXAMPLES.

1. Write 7 tenths; 365 thousandths; 75 millionths.
Ans. 0·7; 0·365; 0·000075.

2 Write 37 hundredths; 5 tenths; 3781 ten millionths.
Ans. 0·37; 0·5; 0·0003781.

3. Write 43 hundredths; 3456 ten thousandths.
Ans. 0·43; 0·3456.

4. Write 13 billionths; 3 ten billionths.
Ans. 0·000000013; 0·0000000003.

---

35. Since decimals, like whole numbers, decrease from the left towards the right, in a ten-fold ratio, they may, when connected together by means of the decimal point, be operated upon by precisely the same rules as for whole numbers, provided we are careful to keep the decimal point always in the right place.

Annexing a cipher to a decimal, does not change its value. Thus, 0·3=0·30=0·300=, &c. But prefixing a cipher is the same as removing the decimal figures one place further to the right, and, therefore, each cipher thus prefixed reduces the value in a ten-fold ratio.

Thus, 0·3 is ten times 0·03, or a hundred times 0·003.

ADDITION OF DECIMALS.

36. FROM what has been said under Art. **35**, we deduce the following

RULE.

Place the numbers so that the decimal points shall be directly over each other, and then add as in whole numbers.

EXAMPLES.

1. Find the sum of 47·3; 37·672; 1·789101; 88·9134, and 0·0037.

Operation.

```
      47·3
      37·672
       1·789101
      88·9134
       0·0037
Ans. 175·678201.
```

2. What is the sum of 0·67; 0·0371; 1100·0001; 47·5; 29·0037; 1·000005, and 33·033?

Ans. 1211·243905.

3. What is the sum of 1·8; 40·06; 120·365; 47·003; 1100·0001; 31·11101, and 3·0001?

Ans. 1343·33921.

4. What is the sum of 13·29; 14·2835; 111·117; 4·006; 67·88864, and 496·446? *Ans.* 707·03114.

5. What is the sum of 37·345; 8·26; 19·0005; 7·534; 10·94, and 103·729? *Ans.* 186·8085.

6. What is the sum of 0·90058; 7·634; 3·007956, and 1·1? *Ans.* 12·642536.

7. What is the sum of 47·635; 3·13; 0·003001; 4·5787; 0·40005, and 4112·3789?

Ans. 4168·125651.

8. What is the sum of 17·154; 32·004501; 49·345; 6·4, and 1·0005? *Ans.* 105·904001.

9. What is the sum of 4·996; 38·37; 421·633; 5·65, and 4·29? *Ans.* 474·939.

10. What is the sum of 57·41; 365·0001, and 1·101?

Ans. 423·5111.

11. What is the sum of 2·4999; 47·121212; 0·1, and 411·001? *Ans.* 460·722112.

12. What is the sum of 433·9; 777·5; 67·06, and 35·88? *Ans.* 1314·34.

SUBTRACTION OF DECIMALS.

37. FROM what has been said under Art. **35**, we infer the following

RULE.

Place the smaller number under the larger, so that the decimal point of the one may be directly under that of the other. Then proceed as in subtraction of whole numbers.

EXAMPLES.

1. From 213·5734 subtract 87·657237.

Operation.

```
     213·5734
      87·657237
Ans. 125·916163.
```

2. From 385·76943 subtract 72·57. *Ans.* 313·19943.

3. From 0·975 subtract 0·483764. *Ans.* 0·491236.

4. From 0·5 subtract 0·0003. *Ans.* 0·4997.

5. From 96·5 subtract 0·000783. *Ans.* 96·499217.

6. From 23·005 subtract 13·000378. *Ans.* 10·004622.

7. From 110·001001 subtract 11·010002.

Ans. 98·990999.

MULTIPLICATION OF DECIMALS.

38. Let us multiply 0·47 by 0·6. If we put these decimals in the form of vulgar fractions, they will become $\frac{47}{100}$, and $\frac{6}{10}$; these, multiplied by Rule under Art. **25**, give $\frac{47}{100} \times \frac{6}{10} = \frac{282}{1000}$. Now it is obvious that there will be, in all cases, as many ciphers in the denominator of the product as there are in the denominators of both the factors added together.

Hence, the following

RULE.

Multiply the two factors after the same manner as in whole numbers; then point off, from the right of the product, as many figures for decimals as there are decimal places in both the factors. If there are not so many places of figures, supply the deficiency by prefixing ciphers.

EXAMPLES.

1. Multiply 3·753 by 1·656.

Operation.

```
       3·753
       1·656
       -----
       22518
      18765
     22518
     3753
    --------
Ans. 6·214968.
```

2. What is the product of 0·005 into 0·017?

Ans. 0·000085.

3. What is the product of 0·376 into 0·0076894?

Ans. 0·0028912144.

4. What is the product of 0·576 into 0·3854?

Ans. 0·2219904.

5. What is the product of 0·43 into 0·65?

Ans. 0·2795.

6. What is the product of 3·9765 into 4·378?

Ans. 17·409117.

7. What is the product of 415·314 into 7·3004?

Ans. 3031·9583256.

8. What is the product of 7·42 into 11·1415?

Ans. 82·66993.

ABRIDGED MULTIPLICATION OF DECIMALS.

39. Abridged multiplication may be advantageously employed, when one or both of the factors are expressed approximately in decimals.

Suppose we wish the product of $\frac{1}{3}$ and $\frac{1}{6}$; if we employ the rule under Art. **25**, we shall find $\frac{1}{3}\times\frac{1}{6}=\frac{1}{18}$.

In decimals, we have $\frac{1}{3}=0·33333$, &c.; $\frac{1}{6}=0·16666$, &c., and $\frac{1}{18}=0·05555$, &c.

We will now multiply together the decimal values of $\frac{1}{3}$, and $\frac{1}{6}$, employing, in the first operation, 3 decimal places in each factor; 4 places in the second operation, and 5 in the third, as follows:

First Operation.	*Second Operation.*	*Third Operation.*
0·333	0·3333	0·33333
0·166	0·1666	0·16666
1 998	1 9998	1 99998
19 98	19 998	19 9998
33 3	199 98	199 998
0·055,278	333 3	1999 98
	0·0555,2778	3333 3
		0·05555,27778

In the first operation, the result is true to only 3 places of decimals; in the second, it is true to 4, and in the third, to 5; and were we to employ a greater number of decimals in each factor, the product would be found to be accurate to only as many decimal places as there were in each factor. And in all cases, when the factors are approximate decimals, the whole number of decimals obtained in the product, by the usual method of multiplication, is not accurate.

By the following rule, we may very much abridge the labor of multiplying, and still obtain the product with the same degree of accuracy as by the usual rule.

Our rule for contracting the work of multiplying decimals, is as follows:

RULE.

I. *Multiply the multiplicand by the left-hand figure of the multiplier.*

II *Multiply the multiplicand, deprived of its right-hand figure, by the second figure of the multiplier, counting from the left.*

III. *Multiply the multiplicand, deprived of its two*

right-hand figures, by the third figure of the multiplier, counting from the left.

Continue this process until all the figures of the multiplier have been used. Observe to place the successive products so that their right-hand figures shall be directly under each other.

NOTE.—In omitting successively the different figures on the right of the multiplicand, we must so far use them as to determine what there would be to carry into the next column.

The student may, perhaps, find some difficulty in fixing the decimal point in the right place. Whenever he is at a loss in this respect, he can multiply a few of the left-hand figures of each factor by the common method, by which means he will be enabled to determine the true place for the decimal point.

Or, which perhaps would be more simple, let the decimal point be fixed in the first partial product, which may be done by the usual rules for decimals.

EXAMPLES.

1. Multiply 0·37894 by 0·67452.

Operation.

```
        0·37894
        0·67452
       --------
       0·227364
          26526
           1516
            189
              7
       --------
Ans.   0·255602
```

Explanation.

First. We multiply the multiplicand 0·37894 by 6, the left-hand figure of the multiplier, which gives the first partial product, 227364

Secondly. We multiply 0·3789, which is the multiplicand deprived of its right-hand figure, by 7, the second figure of the multiplier, observing to carry 3, since the figure cut off, multiplied by 7, gives 28, which is nearer 30 than 20; we thus obtain 26526 for the second partial product.

Thirdly. Multiplying 0·378 by 4, observing to carry 4, we obtain 1516 for the third partial product.

Fourthly. Multiplying 0·37 by 5, observing to carry 4, we obtain 189 for the fourth partial product.

Fifthly. Multiplying 0·3 by 2, observing to carry 1, we get 7 for the fifth partial product.

As a second example, we will find the product of $\frac{1}{3}$ by $\frac{1}{6}$, using in the first operation, 3 decimal places in each factor; in the second operation, we will use 4, and in the third, we will use 5, as follows:

First Operation.	*Second Operation.*	*Third Operation.*
0·333	0·3333	0·33333
0·166	0·1666	0·16666
0·0333	0·03333	0·033333
200	2000	20000
20	200	2000
0·0553	20	200
	0·05553	20
		0·055553

From the above work, it will be seen that the results of these three operations have the same degree of accuracy as when performed by the usual rule.

3. Multiply 0·3785 by 0·4673.

Operation.

```
      0·3785
      0·4673
     -------
     0·15140
        2271
         265
          11
     -------
Ans. 0·17687
```

4. Multiply 0·00524486 by 0·99993682.

Operation.

```
      0·00524486
      0·99993682
     -----------
     0·004720374
          472037
           47204
            4720
             157
              31
               4
     -----------
Ans. 0·005244527
```

5. Multiply 108·2808251671 by 1·9614591767.

Operation.

```
     108·2808251671
       1·9614591767
     --------------
     108·2808251671
      97 4527426504
       6 4968495100
         1082808252
          433123301
           54140412
            9745274
             108281
              75796
               6497
                758
     --------------
Ans. 212·3884181846
```

6. Multiply 0·009416517988 by 0·999936883996.
Ans. 0·0094159236548.

7. Multiply 0·0000375229 by 0·0000275177.
Ans. 0·000000001032543.

8. Multiply 0·999936883996 by 0·999955663612.
Ans. 0·9998925504063.

9. Multiply 0·587401052 by 0·018468950.
Ans. 0·0108486807.

10. Multiply 91·6264232009 by 0·0172021234.
Ans. 1·576169038601.

11. Multiply 212·3880258928 into itself.
Ans. 45108·67354264.

DIVISION OF DECIMALS.

40. In multiplication, we have seen that there are as many decimal places in the product as there are in both the factors; and, since division is the reverse of multiplication, it follows that the number of decimal places in the quotient must equal the excess of those in the dividend, above those of the divisor. Hence, to divide one decimal expression by another, we have this

RULE.

Divide as in whole numbers, and point off as many places from the right of the quotient, for decimals, as the decimal places in the dividend exceed those of the divisor. If there are not as many figures in the quotient as this excess, supply the deficiency by prefixing ciphers.

EXAMPLES.

1. Divide 3·475 by 4·789.

Operation.

```
4·789)3·475000(0·725
      3 3523
      ------
       12270
        9578
       -----
        26920
        23945
        -----
         2975
```

In this example, the number of decimal places in the dividend, including the ciphers which were annexed, is 6, whilst the number of places in the divisor is 3; therefore, we make 3 places of decimals in the quotient. We might continue to annex ciphers to the remainder, and thus obtain additional decimal figures.

2. What is the quotient of 78·56453 divided by 4·78? *Ans.* 16·436.

3. What is the quotient of 1561·275 divided by 24·3? *Ans.* 64·25.

4. What is the quotient of 0·264 divided by 0·2? *Ans.* 1·32.

5. What is the quotient of 3·52275 divided by 3·355? *Ans.* 1·05.

6. What is the quotient of 901·125 divided by 2·25? *Ans.* 400·5.

ABRIDGED DIVISION OF DECIMALS.

41. If we divide 0·30679006 by 0·27610603, by the last rule, our work will be as follows:

Operation.

```
0·27610603)0·30679006(1·1111313
           27610603
            3068403|0
            2761060|3
             307342|70
             276106|03
              31236|670
              27610|603
               3626|0670
               2761|0603
                865|00670
                828|31809
                 36|688610
                 27|610603
                  9|0780070
                  8|2831809
                   |7948261
```

By simply inspecting the above work, it is obvious that all that part of the work which is on the right of the vertical line can in no way affect the accuracy of our quotient figures. By the following rule, we may perform the work of division so as to exclude all that part of the work on the right of the vertical line, thereby shortening the work, and still obtaining as accurate a result as by the last rule.

To contract the work in the division of decimals, we have this

RULE.

Proceed as in the last rule, until we reach that point of the work where it would be necessary to annex ciphers to the remainder. Then, instead of annexing a cipher to the remainder, omit the right-hand figure of the divisor, and we shall obtain the next figure of the quotient;

and thus continue, at each successive figure of the quotient, to omit the right-hand figure of the divisor, until there is but one figure in the remainder.

NOTE.—If we regard the dividend as the numerator of a fraction whose denominator is the divisor, the quotient will be the value of such fraction. Annexing a cipher to the numerator of a fraction is equivalent to multiplying its value by 10, and omitting the right-hand figure of the denominator is also equivalent to multiplying the value of the fraction by 10. Hence, in the operation of division of decimals, instead of annexing a cipher to the dividend, as in the ordinary rules, we may, instead thereof, omit the right-hand figure of the divisor, as in the foregoing rule.

EXAMPLES.

1. What is the quotient of 365·424907 divided by 0·263803?

Operation.

```
0·263803)365·424907(1385·21892
         263 803
         -------
         101 6219
          79 1409
         --------
          22 48100
          21 10424
          --------
           1 376767
           1 319015
           --------
              57752
              52761
              -----
               4991
               2638
               ----
               2353
               2110
               ----
                243
                237
                ---
                  6
                  5
                  -
                  1
```

2. What is the quotient of 0·123456 divided by 1·912478?

Operation.

```
1·912478)0·123456(0·064552
           114748
           ------
             8708
             7650
             ----
             1058
              956
             ----
              102
               96
              ---
                6
                4
                -
                2
```

3. What is the quotient of 0·52600000 divided by 0·5260202?

Operation.

```
0·5260202)0·52600000(0·9999616
            47341818
            --------
             5258182
             4734182
             -------
              524000
              473418
              ------
               50582
               47342
               -----
                3240
                3156
                ----
                  84
                  53
                  --
                  31
                  31
                  --
                   0
```

4. What is the quotient of 7·45678 divided by 4·56789? *Ans.* 1·63243.

5. What is the quotient of 7·632038 divided by 3·716048? *Ans.* 2·053805.

6. What is the quotient of 2 divided by 15·314865? *Ans.* 0·13059207.

7. What is the quotient of 0·926954 divided by 0·3547898? *Ans.* 2·612685.

8. What is the quotient of 13·75892 divided by 6·76897? *Ans.* 2·03264.

---

42. To change a vulgar fraction into an equivalent decimal fraction.

It is obvious that the rule under Art. **33** will apply to this case by considering all the denominate values as decreasing regularly in a ten-fold ratio. Hence, this

RULE.

Annex a cipher to the numerator, and then divide by the denominator; to the remainder annex another cipher, and again divide by the denominator, and so continue, until there is no remainder, or until we have obtained as many decimal figures as may be desired. The successive quotients will be the successive decimal figures required.

EXAMPLES.

1. What decimal fraction is equivalent to $\frac{1}{16}$?

Operation.

```
16)100(0·0625
   96
   --
    40
    32
    --
     80
     80
     --
      0
```

2. What decimal is equivalent to $\frac{1}{18}$?
Ans. 0·05555, &c.
3. What decimal is equivalent to $\frac{1}{20}$? *Ans.* 0·05.
4. What decimal is equivalent to $\frac{1}{25}$? *Ans.* 0·04.
5. What decimal is equivalent to $\frac{1}{3}$?
Ans. 0·3333, &c.
6. What decimal is equivalent to $\frac{6}{125}$?
Ans. 0·048.
7. What decimal is equivalent to $\frac{7}{8}$? *Ans.* 0·875.
8. What decimal is equivalent to $\frac{3}{4}$? *Ans.* 0·75.

Since, in the above process of decimating a vulgar fraction, each successive dividend terminates with a zero, it follows that the right-hand figure of the remainder may be found by multiplying the right-hand figure of the denominator of the vulgar fraction by the quotient figure, and subtracting the right-hand figure of the product from 10; or, which is the same thing, if we subtract the right-hand figure of the denominator from 10, and multiply the remainder by any decimal figure, the right-hand figure of the product will be the same as the right-hand figure of the remainder.

---

43. It will often happen, as in examples 2 and 5, of the last article, that the process will never terminate, in which case there is no decimal value which is accurately equal to the vulgar fraction.

Since we constantly multiplied the remainders by 10, it follows that whenever the denominator of the vulgar fraction contains no prime factors different from those which compose 10, viz., 2 and 5, then the decimal value will *terminate*. But, in all other cases, the decimal expression must consist of an infinite number of figures.

Hence, to determine whether a given vulgar fraction can be accurately expressed in decimals, we have this

RULE.

Decompose the denominator of the vulgar fraction, when reduced to its lowest terms, into its prime factors, (by Rule under Art. **7**,*) then, if there are no prime factors different from* 2 *and* 5, *the vulgar fraction can be accurately expressed by decimals; but if it contain different factors, it cannot be expressed in decimals.*

EXAMPLES.

1. Can the vulgar fraction $\frac{3}{386}$ be accurately expressed in decimals?

In this example, we find that $386=2\times193$; so that the denominator contains the prime factor, 193, which is different from 2 or 5; consequently, $\frac{3}{386}$ cannot be accurately expressed in decimals.

2. Can the vulgar fraction $\frac{17}{812}$ be accurately expressed in decimals? *Ans. It cannot.*

3. Can the vulgar fractions, having for denominators 640, be expressed in decimals accurately?

Ans. They can.

---

44. When a vulgar fraction can be accurately expressed in decimals, we may determine the number of decimal places by the following

RULE.

Decompose the denominator, after the fraction is reduced to its lowest terms, into its prime factors, (*by rule under Art.* **7**,) *which factors cannot differ from* 2 *and* 5, (*by rule under Art.* **43.**) *The highest exponent of* 2, *or* 5, *will be the number of decimal places sought.*

EXAMPLES.

1. How many places of decimals will be required to express $\frac{1}{40}$?

In this example, we find $40=2^3\times 5$, where the highest exponent is 3; therefore, the number of decimal places is 3.

2. How many places of decimals will be required to express $\frac{7}{125}$? *Ans.* 3.

3. How many places of decimals will be required to express $\frac{463}{3125}$? *Ans.* 5.

4. How many places of decimals will be required to express $\frac{13}{16}$? *Ans.* 4.

5. How many places of decimals will be required to express $\frac{37}{160}$? *Ans.* 5.

6. How many places of decimals will be required to express $\frac{43}{1600}$? *Ans.* 6.

---

45. When many figures in the decimal are required, we may proceed as follows:

Required the decimal value of $\frac{1}{29}$?

Following the rule under Art. **42**, we get this

Operation.

```
29)100(0·03448
    87
    ---
    130
    116
    ---
     140
     116
     ---
      240
      232
      ---
        8
```

We have continued this process until we have found a remainder consisting of but one figure; placing this remainder, when divided by 29, at the right of the quotient, agreeably to the usual rules of division, we get,

I. $\frac{1}{29}=0·03448\frac{8}{29}$. Multiplying this by 8, we get $\frac{8}{29}=0·27586\frac{6}{29}$. Substituting this value of $\frac{8}{29}$ in I., we get,

II. $\frac{1}{29}=0·0344827586\frac{6}{29}$; this, multiplied by 6, gives $\frac{6}{29}=0·2068965517\frac{7}{29}$; which, substituted in II., gives,

III. $\frac{1}{29}=0·03448275862068965517\frac{7}{29}$. Again multiplying by 7, we get $\frac{7}{29}=0.24137931034482758620\frac{20}{29}$. Substituting this in III., we get,

IV. $\frac{1}{29}=0·03448275862068965517241379310344827586 20\frac{20}{29}$.

In the expression $\frac{1}{29}=0·03448\frac{8}{29}$, the numerator 8, of the vulgar fraction $\frac{8}{29}$, is the fifth remainder; and in the expression $\frac{1}{29}=0·0344827586\frac{6}{29}$, the numerator 6, of the fraction $\frac{6}{29}$, is the tenth remainder; but this remainder 6, was obtained by multiplying the 5th remainder, which is 8, into itself, and dividing the product, 64, by 29, we thus found the remainder, 6. Again, the 20th remainder, which is 7, was found by multiplying the 10th remainder into itself, and dividing the product by 29. For a similar reason, the remainder of the product of any two remainders will give the remainder corresponding with the sum of the numbers denoting their order; thus, the 5th multiplied by the 7th, will give the 12th; the 6th multiplied by the 9th will give the 15th, and so on for other combinations.

---

46. There is another way of decimating, which is as follows:

Decimate $\frac{1}{97}$.

According to rule under Art. **42**, we find,

```
97)100(0·01
    97
    --
     3
```

To continue this process, we must add ciphers to this remainder in the same way as we did to the numerator, 1. Now, the remainder being 3 times as large as the first numerator, it follows that the next two decimal figures must be 3 times the two just obtained, that is, $3\times01=03$; and, for a similar reason, we must multiply 03 by 3, to obtain the next two figures, and so on.

Proceeding in this way, we find,

```
1/97=0·0103092781
               243
                 729
                  2187
                    6561
                      &c.
-----------------------
1/97=0·010309278350515, &c.
```

Decimating $\frac{1}{4}$ by the above plan, we get

$$\begin{array}{l} \frac{1}{4}=0\cdot248 \\ \quad\;\;\;\;16 \\ \quad\;\;\;\;\;32 \\ \quad\;\;\;\;\;\;64 \\ \quad\;\;\;\;\;\;128 \\ \quad\;\;\;\;\;\;\;256 \\ \quad\;\;\;\;\;\;\;\&c. \\ \hline \end{array}$$

$\frac{1}{4}=0\cdot249999$, &c., which will constantly approximate towards $0\cdot25$; hence, $\frac{1}{4}=0\cdot25$.

Decimating $\frac{1}{8}$ by this method, we get,

$$\begin{array}{l} \frac{1}{8}=0\cdot1248 \\ \quad\;\;\;\;\;16 \\ \quad\;\;\;\;\;\;32 \\ \quad\;\;\;\;\;\;\;64 \\ \quad\;\;\;\;\;\;\;128 \\ \quad\;\;\;\;\;\;\;\;256 \\ \quad\;\;\;\;\;\;\;\;\&c. \\ \hline \end{array}$$

$\frac{1}{8}=0\cdot1249999$, &c., which will constantly approximate towards $0\cdot125$.

---

47. When the decimal figures, obtained by converting a vulgar fraction into decimals, do not terminate, they must recur in periods, whose number of terms cannot exceed the number of units in the denominator, less one. For, all the different remainders which occur must be less than the denominator; and, therefore, their number cannot exceed the denominator, less one; and, whenever we obtain a remainder like one that has previously occurred, then the decimal figures will begin to repeat. Decimals which recur in this way are called *repetends*.

When the period begins with the first decimal figure, it is called a *simple repetend.* But when other decimal figures occur before the period commences, it is called a *compound repetend.*

A repetend is distinguished from ordinary decimals by a period, or dot, placed over the first and last figure of the circulating period.

---

48. The following vulgar fractions give simple repetends:

$\frac{1}{3}=0{\cdot}\dot{3}.$

$\frac{1}{7}=0{\cdot}\dot{1}4285\dot{7}.$

$\frac{1}{9}=0{\cdot}\dot{1}.$

$\frac{1}{11}=0{\cdot}\dot{0}\dot{9}.$

$\frac{1}{13}=0{\cdot}\dot{0}7692\dot{3}.$

$\frac{1}{17}=0{\cdot}\dot{0}58823529411764\dot{7}.$

$\frac{1}{19}=0{\cdot}\dot{0}5263157894736842\dot{1}.$

$\frac{1}{21}=0{\cdot}\dot{0}4761\dot{9}.$

$\frac{1}{23}=0{\cdot}\dot{0}43478260869565217391\dot{3}.$

---

49. The following ones give compound repetends.

$\frac{1}{6}=0{\cdot}1\dot{6}.$

$\frac{1}{12}=0{\cdot}08\dot{3}.$

$\frac{1}{14}=0{\cdot}0\dot{7}1428\dot{5}.$

$\frac{1}{15}=0{\cdot}0\dot{6}.$

$\frac{1}{18}=0{\cdot}0\dot{5}.$

$\frac{1}{22}=0.0\dot{4}\dot{5}.$

$\frac{1}{24}=0{\cdot}041\dot{6}.$

50. Whenever the prime factors of the denominator of a vulgar fraction contain neither of the factors 2 and 5, the repetend will be simple. But when they contain one or both of the factors 2 and 5, together with other factors, then the repetend will be compound.

---

51. Those simple repetends which have as many terms, less one, as there are units in the denominators of their equivalent vulgar fractions, we shall call *perfect repetends*. The following are all of the perfect repetends, whose denominators are less than 100.

$\frac{1}{7}=0{\cdot}\dot{1}4285\dot{7}.$

$\frac{1}{17}=0{\cdot}\dot{0}58823529411764\dot{7}.$

$\frac{1}{19}=0{\cdot}\dot{0}5263157894736842\dot{1}.$

$\frac{1}{23}=0{\cdot}\dot{0}43478260869565217391\dot{3}.$

$\frac{1}{29}=0{\cdot}\dot{0}34482758620689655172413793\dot{1}.$

$$\frac{1}{47}=\begin{cases}0{\cdot}\dot{0}212765957446808510638\\ 29787234042553191489361\dot{7}.\end{cases}$$

$$\frac{1}{59}=\begin{cases}0{\cdot}\dot{0}169491525423728813559322033\\ 898305084745762711864406779661\dot{1}.\end{cases}$$

$$\frac{1}{61}=\begin{cases}0{\cdot}\dot{0}16393442622950819672131147540\\ 98360655737704918032786885245\dot{9}.\end{cases}$$

$$\frac{1}{97}=\begin{cases}0{\cdot}\dot{0}10309278350515463917525773195\,87\\ 628865979381443298969072164948\,45\\ 360824742268041237113402061855\,6\dot{7}.\end{cases}$$

The value of $\frac{1}{17}$ may be made to assume the following forms:

$$\tfrac{1}{17}=0{\cdot}0\tfrac{10}{17}=0{\cdot}05\tfrac{15}{17}=0{\cdot}058\tfrac{14}{17}=0{\cdot}0588\tfrac{4}{17}=0{\cdot}05882\tfrac{6}{17}$$
$$=0{\cdot}058823\tfrac{9}{17}=0{\cdot}0588235\tfrac{5}{17}=0{\cdot}05882352\tfrac{16}{17}=\&c.,$$

where each successive value is extended one decimal place further than its preceding value. The numerators of the vulgar fractions connected with the above decimal expressions are the successive remainders found in the operation of converting the vulgar fraction $\frac{1}{17}$ into a decimal. (Art. **42.**)

If this process of decimating be continued, it will be found to give a simple repetend, consisting of 16 places of figures; it is, therefore, a *perfect repetend.* We will arrange this repetend by placing above each figure its corresponding remainder, as follows;

$$\tfrac{1}{17}=0{\cdot}\;\overset{10}{0}\,\overset{15}{5}\,\overset{14}{8}\,\overset{4}{8}\,\overset{6}{2}\,\overset{9}{3}\,\overset{5}{5}\,\overset{16}{2}\,\overset{7}{9}\,\overset{2}{4}\,\overset{3}{1}\,\overset{13}{1}\,\overset{11}{7}\,\overset{8}{6}\,\overset{12}{4}\,\overset{1}{7}\;\Big|\;\overset{10}{0}\,\overset{15}{5}\,\overset{14,\,\&c.}{8},\ \&c.$$

If we fix our attention upon a particular remainder, as the fifth, for instance, which is 6, it is evident that the decimals which follow, as 3529, &c., continued to infinity, must express the decimal value of $\frac{6}{17}$; for, had we terminated our division after the fifth decimal figure was obtained, we should have had $\frac{1}{17}=0{\cdot}05882\frac{6}{17}$, where $\frac{6}{17}$ stands instead of the decimal figures which follow 0·05882, so that the decimal figures following the remainder, 6, is equal to $\frac{6}{17}$. In the same way, the decimals which follow any other remainder is the value of the vulgar fraction whose denominator is 17, and whose numerator is said remainder.

We have already said that the decimal figures would commence repeating when a remainder is found like one which has previously occurred, (Art. **47.**) A *perfect* *repetend* has been defined (Art. **51**,) as one whose number of decimal figures is equal to one less than the units in the denominator of the vulgar fraction from which it is derived. Therefore, in converting the vulgar fraction $\frac{1}{17}$ into a perfect repetend, every number, from 1 to 16, inclusive, must appear as a remainder. Let us suppose we have reached that point in the process of decimating, which gives 16 for the remainder; then the decimals which follow, being the value of $\frac{16}{17}$, must, when added to the preceding decimals, the value of $\frac{1}{17}$, make a succession of 9's, as 99999, &c., since $\frac{1}{17}+\frac{16}{17}=1$, which, expressed in decimals, is 0·99999, &c., continued to infinity. Hence, when we have obtained as many figures beyond the remainder 16, as we had before we found this remainder, the decimal figures will begin to repeat. But $\frac{1}{17}$, giving a perfect repetend, must extend to 16 figures before repeating; consequently, 16 will occur as the 8th remainder, or when we have obtained one half the number of decimals in the period, and such must be the case with all perfect repetends.

Therefore, the decimal figures of the first half of the period of a perfect repetend, being added to the figures of the second half, must give 99999, &c.; which, if considered as a decimal, is equivalent to a unit. Hence, also, the remainders of the first half of the period, being added to the remainders of the second half, must make, respectively, 17, since their corresponding decimal values make a unit, which is equivalent to $\frac{17}{17}$.

Whenever the number of figures in the period of a simple repetend, arising from decimating a vulgar fraction, whose numerator is 1, and denominator a prime, is even, the remainder which occurs at the middle of the period will be one less than the denominator of the equivalent vulgar fraction, and the figures of the first half of the period, added to those of the second half, will give 99999, &c.; and their corresponding remainders added, must give the denominator of the equivalent vulgar fraction. Such repetends may be called Complementary Repetends. They, of course, include all perfect repetends, as well as many which are not perfect. The following complementary repetends are not perfect.

$$\tfrac{1}{11}=0\cdot\overset{10}{0}\,\overset{1}{9}.$$

$$\tfrac{1}{13}=0\cdot\overset{10}{0}\,\overset{9}{7}\,\overset{12}{6}\,\overset{3}{9}\,\overset{4}{2}\,\overset{1}{3}.$$

$$\tfrac{1}{73}=0\cdot\overset{10}{0}\,\overset{27}{1}\,\overset{51}{3}\,\overset{72}{6}\,\overset{63}{9}\,\overset{46}{8}\,\overset{22}{6}\,\overset{1}{3}.$$

$$\tfrac{1}{89}=0\cdot\overset{10}{0}\,\overset{11}{1}\,\overset{21}{1}\,\overset{32}{2}\,\overset{53}{3}\,\overset{85}{5}\,\overset{49}{9}\,\overset{45}{5}\,\overset{5}{5}\,\overset{50}{0}\,\overset{55}{5}\,\overset{16}{6}\,\overset{71}{1}\,\overset{87}{7}\,\overset{69}{9}\,\overset{67}{7}\,\overset{47}{7}\,\overset{25}{5}\,\overset{72}{2}\,\overset{8}{8}\,\overset{80}{0}\,\overset{88}{8}.\ \Big\}\text{ Half a period.}$$

$$\tfrac{1}{101}=0\cdot\overset{10}{0}\,\overset{100}{0}\,\overset{91}{9}\,\overset{1}{9}.$$

$$\tfrac{1}{103}=0\cdot\overset{10}{0}\,\overset{100}{0}\,\overset{73}{9}\,\overset{9}{7}\,\overset{90}{0}\,\overset{76}{8}\,\overset{39}{7}\,\overset{81}{3}\,\overset{89}{7}\,\overset{66}{8}\,\overset{42}{6}\,\overset{8}{4}\,\overset{80}{0}\,\overset{79}{7}\,\overset{69}{7}\,\overset{72}{6}\,\overset{102}{6}.\ \Big\}\text{ Half a period.}$$

The following vulgar fractions, when decimated, give an even number of figures in a period, and still they are not complementary repetends, their denominators not being primes.

$$\tfrac{1}{21}=0\cdot\dot{0}4761\dot{9}.$$

$$\tfrac{1}{33}=0\cdot\dot{0}\dot{3}.$$

$$\tfrac{1}{39}=0\cdot\dot{0}2564\dot{1}.$$

$$\tfrac{1}{49}=0\cdot\dot{0}204081,\ \text{\&c., to 42 places.}$$

$$\tfrac{1}{51}=0\cdot\dot{0}19607843137254\dot{9}.$$

52. PERFECT REPETENDS possess some very remarkable properties, which we will explain by means of the following figure:

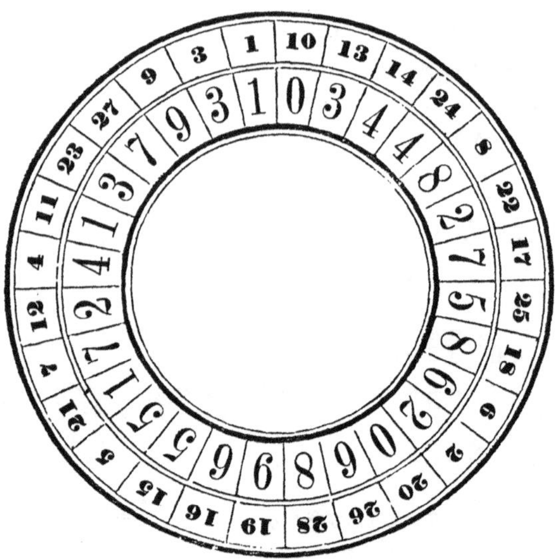

In this figure, the inner circle of figures, commencing at the **0**, directly under the *asterisk*, and counting towards the right hand, is the circulating period of $\frac{1}{29}$.

The outer circle of figures, commencing at the same place, and counting in the same direction, are the successive remainders which will occur in the operation of decimating $\frac{1}{29}$, (by Rule under Art. **42.**)

In this circle of remainders, all the numbers from 1 to 28, inclusive, occur, but not in numerical order.

From what has been said, we infer the following properties, which are common to all *perfect repetends:*

I. *The sum of any two diametrically opposite figures of the circle of decimals, will be* 9.

II. *The sum of any two diametrically opposite terms in the circle of remainders, will make the denominator* 29.

III. *If we subtract the right-hand figure of the denominator from* 10, *and multiply the remainder by any decimal figure of the inner circle, the right-hand figure of the product will be the same as the right-hand figure of the corresponding remainder of the outer circle.*

IV. *Commencing the circle of decimals at any point, and counting completely round, it will be the perfect repetend of the vulgar fraction, whose denominator is the same as in the first case, but whose numerator is the remainder in the outer circle, standing one place to the left.*

V. *If we divide the product of any two remainders by* 29, *what remains will be the remainder in the outer circle, corresponding with the place denoted by the sum of the places of the two numbers.*

From the IVth property, it follows that this same circle of decimals expresses the decimal value of all proper vulgar fractions, whose denominators are 29.

The following figure, formed from the *perfect repetend* of the value of $\frac{1}{23}$, possesses similar properties to those just explained.

Similar circles may be formed for all *perfect repetends.*

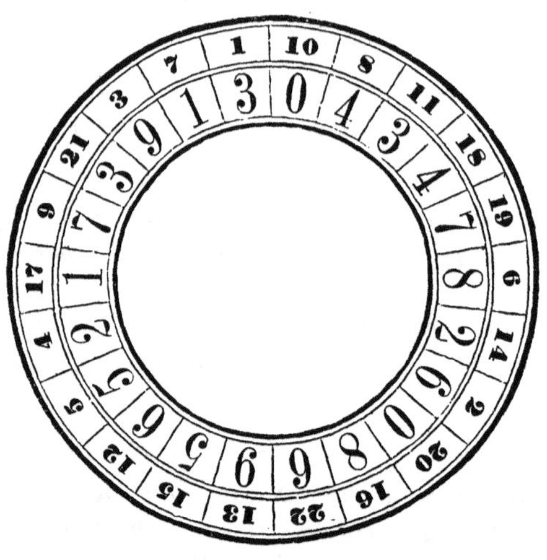

---

53. If we take the *perfect repetend* arising from $\frac{1}{7}$, we find

$\frac{1}{7}=0\cdot\dot{1}4285\dot{7}$. $\frac{4}{7}=0\cdot\dot{5}7142\dot{8}$.

$\frac{2}{7}=0\cdot\dot{2}8571\dot{4}$. $\frac{5}{7}=0\cdot\dot{7}1428\dot{5}$.

$\frac{3}{7}=0\cdot\dot{4}2857\dot{1}$. $\frac{6}{7}=0\cdot\dot{8}5714\dot{2}$.

Taking the *perfect repetend* arising from $\frac{1}{17}$, we find

$\frac{1}{17}=0\cdot\dot{0}58823529411764\dot{7}$.

$\frac{2}{17}=0\cdot\dot{1}17647058823529\dot{4}$.

$\frac{3}{17}=0\cdot\dot{1}76470588235294\dot{1}$.

$\frac{4}{17}=0\cdot\dot{2}35294117647058\dot{8}$.

$\frac{5}{17}=0\cdot\dot{2}94117647058823\dot{5}$.

$\frac{6}{17}=0\cdot\dot{3}52941176470588\dot{2}$.

$\frac{7}{17}=0\cdot\dot{4}11764705882352\dot{9}$.

$\frac{8}{17}=0\cdot\dot{4}70588235294117\dot{6}$.

$\frac{9}{17}=0\cdot\dot{5}29411764705882\dot{3}.$

$\frac{10}{17}=0\cdot\dot{5}88235294117647\dot{0}.$

$\frac{11}{17}=0\cdot\dot{6}47058823529411\dot{7}.$

$\frac{12}{17}=0\cdot\dot{7}05882352941176\dot{4}.$

$\frac{13}{17}=0\cdot\dot{7}64705882352941\dot{1}.$

$\frac{14}{17}=0\cdot\dot{8}23529411764705\dot{8}.$

$\frac{15}{17}=0\cdot\dot{8}82352941176470\dot{5}.$

$\frac{16}{17}=0\cdot\dot{9}41176470588235\dot{2}.$

We will arrange the *complementary repetend* arising from the vulgar fraction $\frac{1}{211}$, in the form of a circle, as was done for *perfect* repetends, as follows:

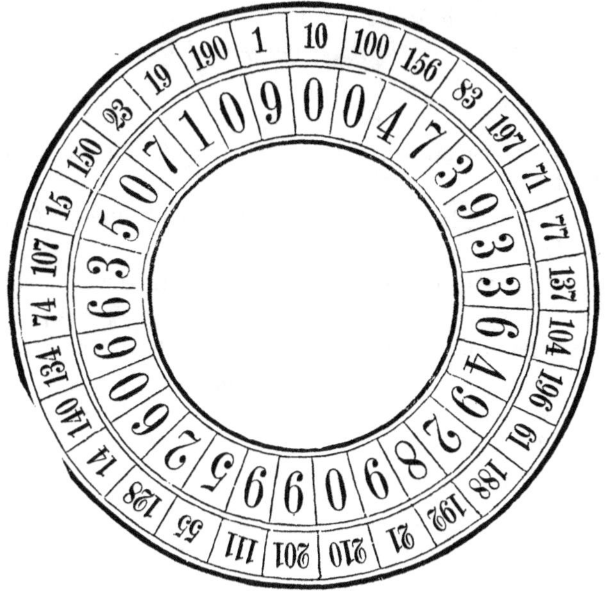

It will be seen that a complementary repetend possesses all the properties ascribed to the perfect repetend, as given under Art. **52**, except the IV.

---

54. To change a decimal fraction into an equivalent vulgar fraction.

Case I.

When the number of places is *finite*, we can, from the definition of decimal fractions, Art. **34**, deduce this

RULE.

Make the given decimal the numerator of the vulgar fraction, and, for its denominator, write 1, *with as many ciphers annexed as there are decimal places.*

EXAMPLES.

1. What vulgar fraction is equivalent to the decimal 0·0625?

$\frac{0625}{10000}$, or $\frac{625}{10000}$; this, reduced by Rule under Art. **17**, gives $\frac{1}{16}$; therefore, $0·0625 = \frac{1}{16}$.

2. What vulgar fraction is equivalent to the decimal 0·134? *Ans.* $\frac{134}{1000} = \frac{67}{500}$.

3. What vulgar fraction is equivalent to the decimal 0·00125? *Ans.* $\frac{125}{100000} = \frac{1}{800}$.

4. What vulgar fraction is equivalent to the decimal 0·0256? *Ans.* $\frac{256}{10000} = \frac{16}{625}$.

5. What vulgar fraction is equivalent to the decimal 0·06248? *Ans.* $\frac{6248}{100000} = \frac{781}{12500}$.

6. What vulgar fraction is equivalent to the decimal 0·001069? *Ans.* $\frac{1069}{1000000}$.

Case II.

When the decimal is a *simple repetend.*

Since $\frac{1}{9}=0\cdot\dot{1}$, it follows that $0\cdot\dot{2}$ must $=\frac{2}{9}$, $0\cdot\dot{3}=\frac{3}{9}$, $0\cdot\dot{4}=\frac{4}{9}$, and so on; therefore, a simple repetend of one figure is equivalent to the vulgar fraction whose numerator is this figure, and whose denominator is 9.

Again, $\frac{1}{99}=0\cdot\dot{0}\dot{1}$; consequently, $0\cdot\dot{0}\dot{7}=\frac{7}{99}$, $0\cdot\dot{4}\dot{5}=\frac{45}{99}$, and so on for other simple repetends of two places of figures.

In a similar manner, we infer that $0\cdot\dot{4}3\dot{2}=\frac{432}{999}$. Therefore, we have the following

RULE.

Make the repetend the numerator; and, for the denominator, write as many nines as there are places of decimals.

EXAMPLES.

1. What vulgar fraction is equivalent to $0\cdot\dot{7}\dot{2}$?

$\frac{72}{99}$; this, reduced by Rule under Art. **17**, becomes $\frac{8}{11}$.

2. What vulgar fraction is equivalent to $0\cdot\dot{1}2\dot{3}$?

Ans. $\frac{123}{999}=\frac{41}{333}$.

3. What vulgar fraction is equivalent to the repetend $0\cdot\dot{0}2\dot{7}$? *Ans.* $\frac{27}{999}=\frac{1}{37}$.

4. What vulgar fraction is equivalent to the repetend $0\cdot\dot{1}4285\dot{7}$? *Ans.* $\frac{142857}{999999}=\frac{1}{7}$.

5. What vulgar fraction is equivalent to the repetend $0\cdot\dot{0}1234567\dot{9}$? *Ans.* $\frac{12345679}{999999999}=\frac{1}{81}$.

6. What vulgar fraction is equivalent to the repetend $0\cdot\dot{0}12345678\dot{9}$? *Ans.* $\frac{123456789}{9999999999}=\frac{13717421}{1111111111}$.

7. What vulgär fraction is equivalent to the repetend $0{\cdot}\dot{1}2332\dot{1}$? *Ans.* $\frac{123321}{999999}=\frac{101}{819}$.

8. What is the value of $0{\cdot}\dot{9}99$ continued to infinity? *Ans.* $\frac{9}{9}=1$.

9. What is the value of $0{\cdot}\dot{9}8765432\dot{0}$? *Ans.* $\frac{987654320}{999999999}=\frac{80}{81}$.

[A very simple method of finding a vulgar fraction equivalent to any repetend, may be found in my ELEMENTS OF ALGEBRA.]

Case III.

When the decimal is a *compound repetend.*

In this case, we obviously have the following

RULE.

I. *Find the vulgar fraction which is equivalent to the decimal figures which precede those that circulate, by Rule under Case I. of this article.*

II. *Find the vulgar fraction which is equivalent to the circulating part of the decimal, by Rule under Case II of this article; to the denominator of this fraction annex as many ciphers as there are decimals which precede the circulating part of the repetend; then add these two fractions together.*

EXAMPLES.

1. What vulgar fraction is equivalent to the compound repetend $0{\cdot}34\dot{3}$? *Ans.* $\frac{34}{100}+\frac{3}{900}=\frac{309}{900}=\frac{103}{300}$.

2. What vulgar fraction is equivalent to the compound repetend $0{\cdot}08\dot{7}8\dot{3}$? *Ans.* $\frac{8}{100}+\frac{783}{99900}=\frac{13}{148}$.

3. What vulgar fraction is equivalent to $0{\cdot}08\dot{3}$? *Ans.* $\frac{8}{100}+\frac{3}{900}=\frac{1}{12}$.

4. What vulgar fraction is equivalent to the compound repetend $0{\cdot}03\dot{5}7142\dot{8}$? *Ans.* $\frac{3}{100}+\frac{571428}{99999900}=\frac{1}{28}$.

5. What vulgar fraction is equivalent to the compound repetend $0{\cdot}0\dot{7}1428\dot{5}$? *Ans.* $\frac{714285}{9999990}=\frac{1}{14}$.

6. What vulgar fraction is equivalent to the compound repetend $0{\cdot}123\dot{4}5\dot{6}$? *Ans.* $\frac{123}{1000}+\frac{456}{999000}=\frac{41111}{333000}$.

If we take the last example, which is $0{\cdot}123\dot{4}5\dot{6}$, and multiply it by 1000000, it will become $123456{\cdot}\dot{4}5\dot{6}$. Again, if we multiply $0{\cdot}123\dot{4}5\dot{6}$ by 1000, it will become $123{\cdot}\dot{4}5\dot{6}$. The difference of these two results is $123456{\cdot}\dot{4}5\dot{6}-123{\cdot}\dot{4}5\dot{6}=123333$. Now, since $123456{\cdot}\dot{4}5\dot{6}$ was 1000000 times the decimal $0{\cdot}123\dot{4}5\dot{6}$, while $123{\cdot}\dot{4}5\dot{6}$ was 1000 times the same decimal, it follows that 123333 is (1000000 − 1000) times its value; that is, 123333 is 999000 times the value of $0{\cdot}123\dot{4}5\dot{6}$; hence, $0{\cdot}123\dot{4}5\dot{6}=\frac{123333}{999000}=\frac{41111}{333000}$, the same as already found. A similar process may be employed for changing any repetend into an equivalent vulgar fraction.

CHAPTER IV.

CONTINUED FRACTIONS.

55. If we divide both numerator and denominator of the fraction $\frac{351}{965}$ by the numerator, we obtain,

I. $\dfrac{351}{965}=\cfrac{1}{2+\cfrac{263}{351}}$. Again, performing the like operation upon the fraction $\frac{263}{351}$, we find $\dfrac{263}{351}=\cfrac{1}{1+\cfrac{88}{263}}$; this value of $\frac{263}{351}$, substituted in I., we get,

II. $\dfrac{351}{965}=\cfrac{1}{2+\cfrac{1}{1+\cfrac{88}{263}}}$. Again, we find $\dfrac{88}{263}=\cfrac{1}{2+\cfrac{87}{88}}$; which, substituted in II., gives,

III. $\dfrac{351}{965}=\cfrac{1}{2+\cfrac{1}{1+\cfrac{1}{2+\cfrac{87}{88}}}}$. Again, $\dfrac{87}{88}=\cfrac{1}{1+\cfrac{1}{87}}$; this value substituted in III., we finally obtain,

IV. $$\frac{351}{965}=\cfrac{1}{2+\cfrac{1}{1+\cfrac{1}{2+\cfrac{1}{1+\cfrac{1}{87}}}}}.$$

By a similar process, we find that

$$\frac{157}{972}=\cfrac{1}{6+\cfrac{1}{5+\cfrac{1}{4+\cfrac{1}{3+\cfrac{1}{2}}}}}.$$

Such fractions as the above are called *continued fractions.*

In the last example, the parts $\frac{1}{6}$, $\frac{1}{5}$, $\frac{1}{4}$, &c., are called the *first, second, third,* &c., *partial fractions.*

It has been proposed, by some authors, to write continued fractions in the following way, which is more compact:

Thus, the preceding fractions may be written,

$$\frac{351}{965}=\frac{1}{2}+\frac{1}{1}+\frac{1}{2}+\frac{1}{1}+\frac{1}{87}.$$

$$\frac{157}{972}=\frac{1}{6}+\frac{1}{5}+\frac{1}{4}+\frac{1}{3}+\frac{1}{2}.$$

If we seek for the greatest common measure of the numerator and denominator of the first fraction $\frac{351}{965}$, by Rule under Art. **9**, we shall have the following

Operation.

```
351)965(2
    702
    263)351(1
        263
         88)263(2
            176
             87)88(1
                87
                 1)87(87
                   87
                    0
```

Here we discover that the successive quotients are the same as the successive denominators of the partial fractions which compose the continued fraction already drawn from $\frac{351}{965}$.

Hence, to convert a vulgar fraction into a continued fraction, we have this

RULE.

Seek, by Rule under Art. **9**, *the greatest common measure of the numerator and denominator of the given fraction; the reciprocals of the successive quotients will form the partial fractions which constitute the continued fraction required.*

EXAMPLES.

1. Convert $\frac{251}{764}$ into a continued fraction.

Operation.

```
251)764(3
    753
    ---
     11)251(22
         22
         --
          31
          22
          --
           9)11(1
              9
              -
              2)9(4
                8
                -
                1)2(2
                  2
                  -
                  0
```

The partial fractions are $\frac{1}{3}$, $\frac{1}{22}$, $\frac{1}{1}$, $\frac{1}{4}$, $\frac{1}{2}$; therefore, we shall have,

$$\frac{251}{764}=\cfrac{1}{3+\cfrac{1}{22+\cfrac{1}{1+\cfrac{1}{4+\cfrac{1}{2.}}}}}$$

2. What continued fraction is equivalent to $\frac{1769}{5537}$?

Ans. $$\cfrac{1}{3+\cfrac{1}{7+\cfrac{1}{1+\cfrac{1}{2+\cfrac{1}{4+\cfrac{1}{5+\cfrac{1}{1+\cfrac{1}{2.}}}}}}}}$$

3. What continued fraction is equivalent to $\frac{113}{194}$?

$$Ans.\ \cfrac{1}{1+\cfrac{1}{1+\cfrac{1}{2+\cfrac{1}{1+\cfrac{1}{1+\cfrac{1}{7+\cfrac{1}{2}}}}}}}.$$

4. What continued fraction is equivalent to $\frac{68}{251}$?

$$Ans.\ \cfrac{1}{3+\cfrac{1}{1+\cfrac{1}{2+\cfrac{1}{4+\cfrac{1}{5}}}}}.$$

---

56. Let us now endeavor to reverse the foregoing process; that is, let us seek the vulgar fraction which is equivalent to a continued fraction.

If we take the continued fraction $\cfrac{1}{2+\cfrac{1}{3+\cfrac{1}{4+\cfrac{1}{5}}}}$, and omit all but the first partial fraction, its value will become $\frac{1}{2}$.

Again, omitting all but the first and second partial fractions, we find $\cfrac{1}{2+\cfrac{1}{3}}=\frac{3}{7}$.

Again, including one more partial fraction, we obtain

$$\cfrac{1}{2+\cfrac{1}{3+\cfrac{1}{4}}}=\frac{13}{30}.$$

When we include the whole, we find

$$\cfrac{1}{2+\cfrac{1}{3+\cfrac{1}{4+\cfrac{1}{5}}}}=\frac{68}{157}.$$

Our successive values, obtained in this way, are $\frac{1}{2}$, $\frac{3}{7}$, $\frac{13}{30}$, and $\frac{68}{157}$.

These values may be derived in the following manner: Take the first partial fraction for the first value; multiply both numerator and denominator by the denominator of the next partial fraction, and we get $\frac{3}{6}$; if we increase this denominator by 1, it will give the second value, $\frac{3}{7}$. Again, multiplying numerator and denominator by the denominator of the next partial fraction, we get $\frac{12}{28}$; if we increase this numerator by the numerator of the last value, also increase the denominator by the denominator of the last value, we get $\frac{13}{30}$, which is the third value. Again, multiplying both numerator and denominator of this value, by the denominator of the next partial fraction, and to the respective products add the numerator and denominator of the preceding value, we obtain the last value, $\frac{68}{157}$.

This last value is the true value of the continued fraction, whilst the other values are successive approximations.

From what has been said, we derive the following Rule for finding the vulgar fraction equivalent to a continued fraction :

RULE.

Consider the symbol $\frac{0}{1}$ as a fraction ; then write this symbol, and the first partial fraction, for the first two terms of the approximate values. Multiply the numerator and denominator of the second approximate value, by the denominator of the next partial fraction, and to the respective products add the numerator and denominator of the next preceding approximate value, and the result will be the succeeding approximate value. Thus continue to multiply the last approximate value by the denominator of the succeeding partial fraction, and to the products add the numerator and denominator of the preceding approximate value ; the result will be the succeeding approximate value.

EXAMPLES

1\. What vulgar fraction is equivalent to the continued fraction $\cfrac{1}{3+\cfrac{1}{2+\cfrac{1}{5+\cfrac{1}{4+\cfrac{1}{6}}}}}$?

In this example, we find the successive approximate values to be $\frac{0}{1}$, $\frac{1}{3}$, $\frac{2}{7}$, $\frac{11}{38}$, $\frac{46}{159}$, and $\frac{287}{992}$.

2. What are the approximative values of the continued fraction

$$\cfrac{1}{3+\cfrac{1}{1+\cfrac{1}{2+\cfrac{1}{4+\cfrac{1}{5}}}}}\ ?$$

Ans. $\frac{0}{1}$, $\frac{1}{3}$, $\frac{1}{4}$, $\frac{3}{11}$, $\frac{13}{48}$, and $\frac{68}{251}$.

3. What are the approximative values of the continued fraction

$$\cfrac{1}{2+\cfrac{1}{2+\cfrac{1}{2+\cfrac{1}{2}}}}\ ?$$

Ans. $\frac{0}{1}$, $\frac{1}{2}$, $\frac{2}{5}$, $\frac{5}{12}$, and $\frac{12}{29}$.

4. What are the approximative values of the continued fraction

$$\cfrac{1}{1+\cfrac{1}{2+\cfrac{1}{3+\cfrac{1}{4+\cfrac{1}{5+\cfrac{1}{6+\cfrac{1}{7+\cfrac{1}{8+\cfrac{1}{9}}}}}}}}}\ ?$$

Ans. $\frac{0}{1}$, $\frac{1}{1}$, $\frac{2}{3}$, $\frac{7}{10}$, $\frac{30}{43}$, $\frac{157}{225}$, $\frac{972}{1393}$, $\frac{6961}{9976}$, $\frac{56660}{81201}$, $\frac{516901}{740785}$.

5. What are the approximative values of the continued fraction $\cfrac{1}{9+\cfrac{1}{8+\cfrac{1}{7+\cfrac{1}{6+\cfrac{1}{5+\cfrac{1}{4+\cfrac{1}{3+\cfrac{1}{2}}}}}}}}$?

Ans. $\left\{ \frac{0}{1}, \frac{1}{9}, \frac{8}{73}, \frac{57}{520}, \frac{350}{3193}, \frac{1807}{16485}, \frac{7578}{60133}, \frac{24541}{223884}, \frac{56660}{516901}. \right.$

---

57. We will now show the application of the foregoing principles of continued fractions by the solution of several practical questions:

1. Express approximately the fractional part of 24 hours, by which the solar year of 365 days, 5 hours, 48 minutes, and 48 seconds, exceeds 365 days.

5 hours, 48 minutes, 48 seconds = 20928 seconds.

24 hours = 86400 seconds. Therefore, the true value of the fraction required is $\frac{20928}{86400}=\frac{109}{450}$.

Now, converting $\frac{109}{450}$ into a continued fraction, by Rule under Art. **55**, we get $\frac{109}{450}=\cfrac{1}{4+\cfrac{1}{7+\cfrac{1}{1+\cfrac{1}{3+\cfrac{1}{1+\cfrac{1}{2}}}}}}$;

and this, re-converted into its approximative values (by Rule under Art. **56**,) gives $\frac{0}{1}$, $\frac{1}{4}$, $\frac{7}{29}$, $\frac{8}{33}$, $\frac{31}{128}$, $\frac{39}{161}$, $\frac{109}{450}$.

The fraction $\frac{1}{4}$ agrees with the correction introduced into the calendar by JULIUS CÆSAR, by means of *bis-sextile* or *leap-year*.

The fraction $\frac{8}{33}$ is the correction used by the Persian astronomers, who add 8 days in every 33 years, by having 7 regular leap-years, and then deferring the eighth until 5 years.

2. The French *metre* is 39·371 inches. Required the approximative ratio of the English foot to the metre.

In this example, the true ratio is $\frac{12000}{39371}$. Operating upon this fraction, as in the last example, we find some of the first approximate values to be $\frac{1}{3}$, $\frac{3}{10}$, $\frac{4}{13}$, $\frac{7}{23}$, $\frac{25}{82}$, $\frac{32}{105}$.

Hence, the foot is to the *metre* as 3 to 10, nearly; a more correct ratio is as 32 to 105.

3. The old Winchester bushel contains 2150·42 cubic inches, and the new Imperial bushel contains 2218·198 cubic inches. Required some of the approximative ratios of these numbers.

In this case, we find some of the approximations to be $\frac{1}{1}$, $\frac{31}{32}$, $\frac{32}{33}$, $\frac{95}{98}$, $\frac{127}{131}$.

Hence, the Winchester bushel is to the Imperial bushel as 32 to 33, nearly. Now, since in a bushel there are 32 quarts, it follows that the Imperial is a Winchester quart larger than the Winchester bushel, nearly.

4. What are some of the approximative values of the ratio of the diameter of a circle to its circumference?

If we take the value of the circumference of the circle, whose diameter is 1, to 10 decimals, we have the vulgar fraction $\frac{10000000000}{31415926535}$, given to find its approximative values.

Proceeding with this, as in the former examples, we find some of the first approximative values to be $\frac{1}{3}$, $\frac{7}{22}$, $\frac{106}{333}$, $\frac{113}{355}$, &c.

---

58. Continued fractions have been the means of obtaining elegant approximations to the roots of surds.

As an example, let it be required to find the square root of $\frac{1}{2}$; or, what is the same thing, the ratio of the side of the square to its diagonal.

In the first place, we obviously have $\frac{1}{\sqrt{2}}=\cfrac{1}{1+\cfrac{\sqrt{2}-1}{1}}$.

Now, if we multiply the numerator and denominator of the fraction $\frac{\sqrt{2}-1}{1}$ by $\sqrt{2}+1$, it will become,

$$\frac{\sqrt{2}-1}{1}=\frac{1}{\sqrt{2}+1}=\cfrac{1}{2+\cfrac{\sqrt{2}-1}{1}}.$$

Therefore, we have,

$$\frac{1}{\sqrt{2}}=\cfrac{1}{1+\cfrac{1}{2+\cfrac{\sqrt{2}-1}{1}}}.$$

Again, the fraction $\frac{\sqrt{2}-1}{1}$ becomes, as before, $=\cfrac{1}{2+\cfrac{\sqrt{2}-1}{1}}$, and by thus continuing this process, we find $\frac{1}{\sqrt{2}}$ to equal the following infinite continued fraction:

$$\cfrac{1}{1+\cfrac{1}{2+\cfrac{1}{2+\cfrac{1}{2+\cfrac{1}{2+, \&c.}}}}}$$

Some of the first approximative values of this fraction are $\frac{0}{1}$, $\frac{1}{1}$, $\frac{2}{3}$, $\frac{5}{7}$, $\frac{12}{17}$, $\frac{29}{41}$, $\frac{70}{99}$, $\frac{169}{239}$, &c.

---

59. We will conclude this subject by pointing out some of the many remarkable properties which the approximative values of continued fractions possess. We will refer to the values just obtained for the ratio of the side of a square to its diagonal.

I. *These values are alternately too small and too large. Thus,* $\frac{0}{1}$, $\frac{2}{3}$, $\frac{12}{17}$, *and* $\frac{70}{99}$, *are too small, while* $\frac{1}{1}$, $\frac{5}{7}$, $\frac{29}{41}$, *and* $\frac{169}{239}$, *are too large.*

II. *Any of these values differ from the true value by a quantity which is less than the reciprocal of the square of its denominator. Thus,* $\frac{12}{17}$, *which is the ratio much used by carpenters in cutting braces, differs from the true ratio by a quantity less than* $(\frac{1}{17})^2 = \frac{1}{289}$.

III. *Any two consecutive terms of these approximate values, when reduced to a common denominator, will differ by a unit in their numerators. Thus,* $\frac{5}{7}$, *and* $\frac{12}{17}$, *when reduced to a common denominator, become* $\frac{85}{119}$, *and* $\frac{84}{119}$.

IV. *The numerator and denominator of all approxi-*

mative values of continued fractions are prime to each other; that is, they have no common measure.

[For a more extended developement of the properties of *Continued Fractions*, see my Treatise on ALGEBRA.]

LAMBERT'S METHOD

OF

DE-COMPOUNDING VULGAR FRACTIONS.

60. THERE is another method of de-compounding fractions, which was first given by the celebrated LAMBERT.

As an example, we will resume our fraction $\frac{351}{965}$, which may be made to take these forms:

$$\frac{351}{965}=\frac{2\times 351}{2\times 965}=\frac{702}{2\times 965}=\frac{965-263}{2\times 965}=\frac{1}{2}-\frac{1}{2}\times\frac{263}{965}.$$

Again,

$$\frac{263}{965}=\frac{3\times 263}{3\times 965}=\frac{789}{3\times 965}=\frac{965-176}{3\times 965}=\frac{1}{3}-\frac{1}{3}\times\frac{176}{965}.$$

And,

$$\frac{176}{965}=\frac{5\times 176}{5\times 965}=\frac{880}{5\times 965}=\frac{965-85}{5\times 965}=\frac{1}{5}-\frac{1}{5}\times\frac{85}{965}.$$

And,

$$\frac{85}{965}=\frac{11\times 85}{11\times 965}=\frac{935}{11\times 965}=\frac{965-30}{11\times 965}=\frac{1}{11}-\frac{1}{11}\times\frac{30}{965}.$$

And,

$$\frac{30}{965}=\frac{32\times 30}{32\times 965}=\frac{960}{32\times 965}=\frac{965-5}{32\times 965}=\frac{1}{32}-\frac{1}{32}\times\frac{5}{965}.$$

And,

$$\frac{5}{965}=\frac{1}{193}.$$

Hence, by successive substitutions, we find,

$$\frac{351}{965}=\frac{1}{2}-\frac{1}{2\times3}+\frac{1}{2\times3\times5}-\frac{1}{2\times3\times5\times11}$$
$$+\frac{1}{2\times3\times5\times11\times32}-\frac{1}{2\times3\times5\times11\times32\times193}.$$

Here we observe that the successive terms are alternately *plus* and *minus*. We also see that the successive factors of the different denominators may be found by continually dividing the denominator, 965, by the numerator, 351, and the successive remainders.

To make this more clear, we will give the above method of division at full length.

Operation.

```
351)965(2
    702
    263)965(3
        789
        176)965(5
            880
             85)965(11
                935
                 30)965(32
                    960
                      5)965(193
                        965
                          0
```

As a second example, de-compound the fraction $\frac{1200}{3937}$, which is nearly the ratio of the English foot to the French metre.

Operation.

```
1200)3937(3
     3600
      337)3937(11
          3707
           230)3937(17
               3910
                 27)3937(145
                    3915
                      22)3937(178
                         3916
                           21)3937(187
                              3927
                                10)3937(393
                                   3930
                                      7, &c.
```

Hence, the value of $\frac{1200}{3937}$

$$=\frac{1}{3}-\frac{1}{3.11}+\frac{1}{3.11.17}-\frac{1}{3.11.17.145}+\frac{1}{3.11.17.145.178}$$

$$-\frac{1}{3.11.17.145.178.187}+\frac{1}{3.11.17.145.178.187.393}-\text{\&c.}$$

It is obvious that the terms of this series converge very rapidly. If we use only one term of the series, we have $\frac{1}{3}$ for the ratio. If we use two terms, we find $\frac{10}{33}$ for the ratio; and in this way, we may find the successive approximate ratios.

If we endeavor to de-compound the fraction $\frac{28673}{45360}$ by this method, we shall find it equivalent to this series of fractions:

$$\frac{1}{1}-\frac{1}{1.2}+\frac{1}{1.2.3}-\frac{1}{1.2.3.4}+\frac{1}{1.2.3.4.5}-\frac{1}{1.2.3.4.5.6}$$

$$+\frac{1}{1.2.3.4.5.6.7}-\frac{1}{1.2.3.4.5.6.7.9}.$$

It will be observed that the factors constituting the denominators of these fractions are the successive digits, except that in the last term the digit 8 does not appear. Now, instead of this last term, we may write these two terms

$$-\frac{1}{1.2.3.4.5.6.7.8}+\frac{1}{1.2.3.4.5.6.7.8.9},$$

which are obviously equivalent in value.

This change being made, we have

$$\frac{28673}{45360}=\frac{1}{1}-\frac{1}{1.2}+\frac{1}{1.2.3}-\frac{1}{1.2.3.4}+\frac{1}{1.2.3.4.5}-\frac{1}{1.2.3.4.5.6}$$
$$+\frac{1}{1.2.3.4.5.6.7}-\frac{1}{1.2.3.4.5.6.7.8}+\frac{1}{1.2.3.4.5.6.7.8.9}.$$

If we de-compound the fraction $\frac{18131}{46080}$ by this method, we shall find it equal to

$$\frac{1}{2}-\frac{1}{2.4}+\frac{1}{2.4.6}-\frac{1}{2.4.6.8}+\frac{1}{2.4.6.8.10}-\frac{1}{2.4.6.8.10.12}.$$

CHAPTER V.

RULE OF THREE.

61. THE quotient arising from dividing one quantity by another of the same kind, is called a *ratio.*

Thus, the ratio of 12 to 3, is $12 \div 3 = \frac{12}{3} = 4$.

The ratio of 15 yards to 5 yards, is $\frac{15}{5} = 3$.

So that the ratio of one number to another is nothing more than the value of a vulgar fraction, whose numerator is the first term, and denominator the last term.

The ratio of 7 days to 5 days, is $\frac{7}{5}$.

The ratio of $3\frac{1}{4}$ hours to $7\frac{1}{2}$ hours, is $3\frac{1}{4} \div 7\frac{1}{2} = \frac{13}{30}$.

When there are four quantities, of which the ratio of the first to the second is the same as that of the third to the fourth, these four quantities are said to be in *proportion.* Thus, 4, 6, 8, and 12, are in proportion, since the ratio of 4 to 6, is the same as 8 to 12. That is, $\frac{4}{6} = \frac{8}{12}$.

Hence, a proportion is nothing more than an equality of ratios.

The usual method of denoting that four terms are in proportion, is by means of points, or dots.

Thus, 4 : 6 : : 8 : 12; where two points are placed between the first and second terms, and also between the third and fourth, and four points are placed between the second and third; which is read, 4 is to 6 as 8 is to 12.

The first and fourth terms of a proportion are called the *extremes*. The second and third terms are called the *means*.

The first and second constitute the *first couplet*.

The third and fourth constitute the *second couplet*.

The two terms of a couplet must be of the same name, or kind; since two quantities of different kinds cannot have a ratio. There can be no ratio between yards and dollars; but the numbers which represent the number of yards and dollars may have a ratio.

Since, in a proportion, the quotient of the first term, divided by the second, is equal to the quotient of the third, divided by the fourth, it follows that the product of the extremes is equal to the product of the means.

Hence, if we divide the product of the means by the first term, we shall obtain the fourth term.

This process of finding the fourth term by means of the other three terms, is called the *Rule of Three*.

Suppose it is required to find what 156 yards of cloth will cost, if 22 yards cost $20·90.

Had there been twice as many yards in 156 as in 22, they would obviously cost twice $20·90; were there three times as many yards, their cost would be three times $20·90, and so on for other ratios.

Hence, the $20·90 must be repeated as many times as 22 yards is continued times in 156 yards; that is, $\frac{156}{22}$ times.

So that, if 22 yards of cloth cost $20·90, 156 yards will cost $\frac{156}{22}$ times $20·90 $=\frac{78}{11}$ times $20·90 = 78 times $\frac{1}{11}$ of $20·90 = 78 times $1·90 = $148·20.

Again, suppose 15 francs to equal $14·10, how many francs are there in $34·78?

Since 15 francs is equal to \$14·10, twice \$14·10 would equal twice 15 francs; thrice \$14·10 would equal thrice 15 francs, and so for other ratios. Therefore, 15 francs must be repeated as many times as \$14·10 is contained times in \$34·78; that is, $\frac{3478}{1410}=\frac{37}{15}$ times.

Hence, the number of francs required is $\frac{37}{15}$ of 15 francs=37 times $\frac{1}{15}$ of 15 francs=37 times 1 franc= 37 francs.

The above method of working questions of the *Rule of Three* is contained in the following simple

RULE.

Of the three terms which are given, one will always be of the same kind as the answer sought; this will be the third term. Then if, by the nature of the question, the answer is required to be greater than the third term, divide the greater of the two remaining terms by the less, for a ratio; but if the answer is required to be less than the third term, then divide the less of the two remaining terms by the greater, for a ratio. Having obtained the ratio, multiply the third term by it, and it will give the answer in the same denomination as the third term.

NOTE.—Before obtaining the ratio by means of the first two terms, we must reduce them to like denominations.

EXAMPLES.

1. If in 7 weeks there are 49 days, how many days are there in 21 weeks?

In this example, the answer is required to be in days; therefore, we must take 49 days for our third term. And, since in 21 weeks there must be more days than in 7 weeks, we get our ratio by dividing 21 by 7, which gives $\frac{21}{7}$; the third term, 49 days, multiplied by this, gives 49 days $\times\frac{21}{7}$, which, by canceling, becomes 147 days, for our answer.

2. If a person perform a journey in 20 days, by traveling 10 hours each day, how long would it take him to perform the same if he travels 8 hours each day?

In this example, our answer is required to be in days; therefore, we must take 20 days for our third term. And, since it will evidently take more days when he travels 8 hours each day, than it did when he traveled 10 hours each day, we must divide 10 hours by 8 hours, for our ratio, which becomes $\frac{10}{8}$; 20 days, multiplied by this, gives 20 days $\times\frac{10}{8}$, which, by canceling, becomes 25 days for the answer.

3. If $\frac{11}{13}$ of a pound of sugar cost $\frac{23}{26}$ of a shilling, how much will $\frac{9}{23}$ of a pound cost?

In this example, our third term is $\frac{23}{26}$ of a shilling. And, since $\frac{9}{23}$ of a pound is less than $\frac{11}{13}$, we must obtain our ratio by dividing $\frac{9}{23}$ by $\frac{11}{13}$, which gives $\frac{9}{23}\times\frac{13}{11}$; this, multiplied by the third term, $\frac{23}{26}$ of a shilling, will give $\frac{23}{26}$ of a shilling $\times\frac{9}{23}\times\frac{13}{11}$. To reduce this with the least labor, we must resort to the method of canceling. Thus, canceling the 23, which occurs in both numerator and denominator, also 13 of the numerator against a part of the 26 of the denominator, our expression will, by this means, become $\frac{1}{2}$ of a shilling $\times\frac{9}{1}\times\frac{1}{11}=\frac{9}{22}$ of a shilling.

NOTE.—This method of canceling should be used when the nature of the question will admit, since it will always very much simplify the operation.

4. If $3\frac{1}{2}$ pounds of coffee cost $2\frac{1}{3}$ shillings, how much will $10\frac{1}{6}$ pounds cost?

In this example, $2\frac{1}{3}=\frac{7}{3}$ shillings must be our third term, and, since $10\frac{1}{6}=\frac{61}{6}$ pounds must cost more than $3\frac{1}{2}=\frac{7}{2}$ pounds, we must divide $\frac{61}{6}$ by $\frac{7}{2}$, for the ratio, making it $\frac{61}{6}\times\frac{2}{7}$; this, multiplied by the third term, $\frac{7}{3}$ shillings, will give $\frac{7}{3}$ shillings $\times\frac{61}{6}\times\frac{2}{7}$; this becomes, after canceling, $\frac{1}{3}$ of a shilling $\times\frac{61}{3}=\frac{61}{9}$ shillings $=6\frac{7}{9}$ shillings.

It may happen that the three given quantities are all of one denomination; still it will be found that two of them are of one kind, and the third one of another kind, which is like the answer. As an example: What tax must an income of \$6000 pay, provided \$100 pays \$1·03?

Here the three given quantities are \$6000, \$100, \$1·03, which are all of the same species, viz., money. Nevertheless, the first and second are *incomes*, and the third is a *tax*. So that the ratio must be \$6000 ÷ \$100 $=\frac{6000}{100}=60$; therefore the tax sought is 60 times \$1·03 = \$61·80.

5. If a tree 38 feet 9 inches in height give a shadow of 49 feet 2 inches, how high is that tree, which, at the same time, casts a shadow of 71 feet 7 inches?

In this example, our third term is the height of the first tree, which is 38 feet 9 inches $=38\frac{3}{4}$ feet $=\frac{155}{4}$ feet. Our ratio will be obtained by dividing 71 feet 7 inches $=71\frac{7}{12}$ feet $=\frac{859}{12}$ feet, by 49 feet 2 inches $=49\frac{1}{6}$ feet $=\frac{295}{6}$ feet, which becomes $\frac{859}{12}\times\frac{6}{295}$; this, multiplied by the third term, $\frac{155}{4}$ feet, gives $\frac{155}{4}$ feet $\times\frac{859}{12}\times\frac{6}{295}$. Canceling 6 of the numerator against a part of the 12 of the denominator; also canceling 5, a factor of 155 of the numerator, against 5, a factor of 295 of the denominator, we get $\frac{31}{4}$ feet $\times\frac{859}{2}\times\frac{1}{59}=\frac{26629}{472}=56\frac{197}{472}$ feet, for the answer.

6. If $16\frac{1}{7}$ yards of calico cost $21\frac{1}{3}$ shillings, how much can be bought for $32\frac{3}{7}$ shillings?

Ans. $24\frac{1689}{3136}$ yards.

7. Sold a tankard for £5 6 s. at the rate of 5 s. 4 d. per ounce. What was its weight?

Ans. 1 lb. 7 oz. 17 pwt. 12 gr.

8. If 300 men consume 70 barrels of provisions in 10 months, how much will 240 men consume in the same time? *Ans.* 56 barrels.

9. Gave 72 dollars for 5 barrels of fish. How much will 20 barrels cost at the same rate? *Ans.* \$288.

10. If it take 30 yards of carpeting, which is $\frac{3}{4}$ of a yard wide, to cover a floor, how many yards, which is $1\frac{1}{4}$ yards wide, will it take to cover the same floor?

Ans. 18 yards.

11. If an individual receive a salary of \$700 per year, how much will that be for each day, counting 365 days for a year? *Ans* $\$1\frac{6\cdot7}{73}$.

12. If the circumference of a wheel is $17\frac{1}{2}$ feet, what distance will it pass over in revolving $13\frac{1}{3}$ times?

Ans. $233\frac{1}{3}$ feet.

13. If it take $2\frac{1}{2}$ yards of cloth for 2 pair of pantaloons, and $12\frac{1}{3}$ yards for 5 coats, how many yards will it require for 7 coats and 8 pair of pantaloons?

Ans. $27\frac{4}{15}$ yards.

14. If $19\frac{1}{2}$ yards of cloth cost $13\frac{1}{7}$ dollars, how many dollars will $14\frac{1}{3}$ yards cost? *Ans.* $\$9\frac{541}{819}$.

15. If 3 bushels of apples cost 13 shillings, how much will $17\frac{1}{2}$ bushels cost? *Ans.* £3 15 s. 10 d.

16. How much must be paid for $3\frac{3}{8}$ cords of wood, at $2\frac{1}{2}$ dollars per cord? *Ans.* $\$8\frac{7}{16} = \$8{\cdot}4375$.

17. If a board 16 feet long and 9 inches wide contain

12 square feet, how long must another board be, which is 8 inches wide, to contain the same number of square feet? *Ans.* 18 feet.

18. If the freight of 7 hogsheads of molasses for 18 miles is 9 dollars, what must be paid for the freight of 21 hogsheads the same distance? *Ans.* $27.

19. If $\frac{3}{8}$ of a vessel is worth $1729, how much is the whole vessel worth? *Ans.* $$4610\frac{2}{3}$.

20. Lent my friend $300 for six months; afterwards he lent me $400. How long may I keep it to balance the favor? *Ans.* $4\frac{1}{2}$ months.

21. If a piece of land 20 rods long, and 8 rods wide, make one acre, how wide must it be, when it is 50 rods long, to contain the same? *Ans.* $3\frac{1}{5}$ rods.

22. If a ship sail $90\frac{3}{8}$ miles in $7\frac{1}{2}$ hours, how many hours will it require to sail 60 miles?

Ans. $4\frac{236}{241}$ hours.

23. If $\frac{3}{4}$ of an acre of land is worth 54 dollars, what is $\frac{1}{10}$ of an acre worth? *Ans.* $7·20.

24. If I pay $\frac{3}{4}$ of a dollar for sawing one cord of wood, how much must I pay for sawing $5\frac{3}{7}$ cords?

Ans. $$4\frac{1}{14}$.

25. If $16\frac{1}{2}$ yards of cloth are worth $\frac{2}{3}$ of $\frac{7}{8}$ of $20\frac{1}{2}$ dollars, what is that per yard? *Ans.* $$\frac{287}{396}$.

26. A man is $1700 in debt, and his estate amounts to but $870. How much can he pay on a dollar?

Ans. $0·51$\frac{3}{17}$.

27. How many yards of paper, $\frac{3}{4}$ wide, will paper 375 square yards? *Ans.* 500 yards.

28. If a staff 5 feet long cast a shade $6\frac{1}{2}$ feet, what is the height of that steeple, whose shade at the same time measures 150 feet? *Ans.* $115\frac{5}{13}$ feet.

29. If 3 paces, or common steps of a person, be equal to 2 yards, how many yards are there in 170 paces? *Ans.* $113\frac{1}{3}$ yards.

30. What cost 3 cwt. of coffee at 15 d. per pound? *Ans.* £21.

31. A garrison of 540 men have provisions for 365 days. How long will those provisions last, if the garrison be increased to 1127 men? *Ans.* $174\frac{1002}{1127}$ days.

32. What will be the tax upon £763 13 s. at the rate of 3 s. 4 d. on one pound sterling? *Ans.* £127 5 s. 6 d.

33. What is the value of a year's rent of 547 acres of land, at the rate of 15 s. 6 d. per acre? *Ans.* £423 18 s. 6 d.

34. Allowing the French metre to be $3\frac{7}{25}$ feet in length, how many feet are there in 46 metres?

Ans. $3\frac{7}{25}$ feet $\times \frac{46}{1} = \frac{82}{25}$ feet $\times \frac{46}{1} = 150\frac{22}{25}$ feet.

35. Suppose a certain quantity of hay will feed 70 sheep 31 days, how long would it keep 131 sheep? *Ans.* $16\frac{74}{131}$ days.

36. If 9 yards of silk, which is $\frac{3}{4}$ of a yard wide, line a cloak, how many yards that is $\frac{5}{4}$ wide, will it take to line the same?

Ans. 9 yards $\times \frac{3}{4} \times \frac{4}{5} =$ 9 yards $\times \frac{3}{5} = 5\frac{2}{5}$ yards.

37. If a barrel of beer last 10 men 16 days, how long will it last 27 men? *Ans.* $5\frac{25}{27}$ days.

38. If $9\frac{2}{7}$ barrels of flour are consumed by a company in 18 days, how long will $35\frac{3}{4}$ barrels last?

$$\textit{Ans.}\ 18 \text{ days} \times \frac{35\frac{3}{4}}{9\frac{2}{7}} = 18 \text{ days} \times \frac{143}{4} \times \frac{7}{65} = 69\frac{3}{10} \text{ days.}$$

39. If a mill grind $19\frac{1}{4}$ bushels of corn in 1 hour, 17 minutes, in what time will it grind 100 bushels?

$19\frac{1}{4}=\frac{77}{4}$; 1 hour, 17 minutes=77 minutes. Hence, we have,

Ans. 77 min.$\times\frac{100}{1}\times\frac{4}{77}$=400 min.=6 h. 40 min.

40. If a barrel of flour will support 13 men for 27 days, how long would it support 9 men?

Ans. 39 days.

41. If $\frac{5}{8}$ of an acre of land cost $13, how much can be bought for $39? *Ans.* $1\frac{7}{8}$ acres.

42. If $\frac{3}{16}$ of a dollar will pay for $\frac{3}{4}$ of a bushel of apples, how many bushels can be bought for $7\frac{7}{8}$ dollars?

Ans. $31\frac{1}{2}$ bushels.

43. If 750 barrels of cider cost $2250, how much will 419 barrels cost? *Ans.* $1257.

44. If $\frac{3}{15}$ of a firkin of butter is worth $1·80, what is $\frac{3}{7}$ of a firkin worth? *Ans.* $3·85$\frac{5}{7}$.

45. If a staff 3 feet in length give a shadow 7 feet long, how high is that tree, which, at the same time, casts a shadow of 90 feet? *Ans.* $38\frac{4}{7}$ feet.

46. A regiment of soldiers, consisting of 976 men, is to be clothed, each coat to contain $2\frac{1}{2}$ yards of cloth, $1\frac{5}{8}$ yards wide, and to be lined with shalloon $\frac{7}{8}$ of a yard wide. How many yards of shalloon will be required?

Since each coat is to contain $2\frac{1}{2}=\frac{5}{2}$ yards, the number of yards for the whole regiment will be $976\times\frac{5}{2}$. Our expression will be $\frac{5}{2}$ yards $\times\frac{976}{1}\times1\frac{5}{8}\times\frac{8}{7}=\frac{5}{2}$ yards $\times\frac{976}{1}\times\frac{13}{7}=4531\frac{3}{7}$ yards.

47. A person owning $\frac{3}{5}$ of a coal-mine, sells $\frac{3}{4}$ of his share for $400. What is the whole mine worth at the same rate? *Ans.* 888\frac{8}{9}$.

48. A and B hire a pasture for $50, in which A pastures 13 cows, and B 12. What must each pay?

The whole number of cows pastured is 13+12=25.

The ratio of A's to the whole is $\frac{13}{25}$. The ratio of B's to the whole is $\frac{12}{25}$. Hence, A must pay $\$50\times\frac{13}{25}=\$26$. B must pay $\$50\times\frac{12}{25}=\$24$.

49. Suppose sound to move 1106 feet in a second; how many miles distant is a cloud in which lightning is observed $47\frac{1}{2}$ seconds before the thunder is heard, no allowance being made for the progressive motion of light? *Ans.* $9\frac{1003}{1056}$ miles.

50. If A can mow an acre of grass in $5\frac{2}{3}$ hours, and B can mow $1\frac{1}{3}$ acres in $8\frac{1}{2}$ hours, in what time can they jointly mow $8\frac{3}{4}$ acres?

Since A can mow 1 acre in $5\frac{2}{3}=\frac{17}{3}$ hours, he can mow $\frac{3}{17}=\frac{9}{51}$ of an acre in 1 hour.

And, since B can mow $1\frac{1}{3}=\frac{4}{3}$ acres in $8\frac{1}{2}=\frac{17}{2}$ hours, he can mow $\frac{4}{3}\div\frac{17}{2}=\frac{8}{51}$ of an acre in 1 hour.

A and B can, together, mow $\frac{9}{51}+\frac{8}{51}=\frac{1}{3}$ of an acre in 1 hour. Hence, to mow $8\frac{3}{4}=\frac{35}{4}$ acres, they will require $\frac{35}{4}\div\frac{1}{3}=\frac{35}{4}\times\frac{3}{1}=\frac{105}{4}=26\frac{1}{4}$ hours.

COMPOUND PROPORTION.

62. When the quantity required depends upon more than three terms, the operation of finding it is called the *rule of compound proportion.*

Suppose we have the following example:

If 20 men, working 10 hours each day, have been employed 18 days in constructing 500 feet of railroad, how many days, of 12 hours each, must 76 men be employed to construct 1140 feet of the same road?

Had the number of feet of road, as well as the number of hours each day employed in labor, been the same in both cases, the question would have been equivalent to the following:

If 20 men have been employed 18 days to perform a certain work, how many days would 76 men require to accomplish the same work?

It is evident that the time sought in this case is the same fractional part of 18 days that 20 men is of 76 men; that is, the time required is

$$\tfrac{20}{76} \text{ of } 18 \text{ days.}$$

If, now, we take into account the number of hours employed each day, still supposing the number of feet of road to remain the same in both cases, our question will read thus:

If it require $\frac{20}{76}$ of 18 days to accomplish a certain work, when 10 hours are each day devoted to labor, how many days will be necessary when 12 hours are reckoned each day?

The answer, in this case, is obviously

$$\tfrac{10}{12} \text{ of } \tfrac{20}{76} \text{ of } 18 \text{ days.}$$

Now, taking into the account the number of feet of road, our question will become as follows:

If $\frac{10}{12}$ of $\frac{20}{76}$ of 18 days are required to construct 500 feet of railroad, how many days will be required to con struct 1140 feet?

This leads to the following final result:

$$\tfrac{1140}{500} \text{ of } \tfrac{10}{12} \text{ of } \tfrac{20}{76} \text{ of } 18 \text{ days} = 9 \text{ days.}$$

From the above work we see that questions of compound proportion may be solved by the following

RULE.

Among the terms given, there will be but one like the answer, which we will call the odd term. The other terms will appear in pairs, or couplets. Form ratios out of each couplet in the same manner as in the Rule of Three; then multiply the odd term and all the ratios together, and it will give the answer in the same name and denomination as the odd term.

NOTE.—Before forming ratios from the couplets, they must be reduced to the same denominate value.

EXAMPLES.

1. If a person travel 300 miles in 17 days, traveling only 6 hours each day, how many miles could he have gone in 15 days, by traveling 10 hours each day?

In this example, the answer is required in miles; therefore, our odd term is 300 miles.

The first couplet consists of days; and, since in 15 days, other things being the same, he could not travel as far as in 17 days, we must divide 15 by 17, which gives $\frac{15}{17}$ for the first ratio.

The second couplet consists of hours; and, since in 10 hours he could travel farther than in 6 hours, we must divide 10 by 6, which gives $\frac{10}{6}$ for the second ratio.

Multiplying these two ratios and the odd term together, we get 300 miles $\times \frac{15}{17} \times \frac{10}{6}$. Canceling the 6 of the denominator against a part of 300 of the numerator, it becomes 50 miles $\times \frac{15}{17} \times \frac{10}{1} = 441\frac{3}{17}$ miles, for the answer.

2. If a marble slab, 10 feet long, 3 feet wide, and 3 inches thick, weigh 400 pounds, what will be the weight

of another slab, of the same marble, whose length is 8 feet, width 4 feet, and thickness 5 inches?

In this example, the answer is required to be in pounds; therefore, 400 pounds is the odd term. The first couplet consists of the lengths; and, since 8 feet in length will give less weight than 10 feet, we must divide 8 by 10, which gives $\frac{8}{10}$ for the first ratio.

The second couplet consists of widths; and, since 4 feet wide will give more weight than 3 feet, we must divide 4 by 3, which gives $\frac{4}{3}$ for the second ratio.

The third couplet consists of thicknesses; and, since 5 inches thick will give more weight than 3 inches, we must divide 5 by 3, which gives $\frac{5}{3}$ for the third ratio.

Multiplying the odd term and these ratios together, we get 400 lbs. $\times \frac{8}{10} \times \frac{4}{3} \times \frac{5}{3}$. Canceling the 10 of the denominator against a part of the 400 of the numerator, we get 40 lbs. $\times \frac{8}{1} \times \frac{4}{3} \times \frac{5}{3} = \frac{6400}{9} = 711\frac{1}{9}$ pounds, for the answer.

3. If a pile of wood, 8 feet long, 4 feet wide, and 4 feet high, contain 1 cord, how many cords are there in a pile 26 feet long, 8 feet wide, and 12 feet high?

In this example, 1 cord is the odd term. The first couplet consists of lengths; and, since 26 feet long will give more wood than 8 feet, we shall have $\frac{26}{8}$ for the first ratio.

The second couplet consists of widths; and, since 8 feet wide will give more than 4 feet, we get $\frac{8}{4}$ for the second ratio.

The third couplet consists of heights; and, since 12 feet high will give more than 4 feet, we get $\frac{12}{4}$ for the third ratio.

Multiplying these ratios and the odd term, 1 cord,

together, we get 1 cord $\times \frac{26}{8} \times \frac{8}{4} \times \frac{12}{4}$. Canceling the 8 of the numerator against the 8 of the denominator, also one of the 4's of the denominator against a part of the 12 of the numerator, and the factor 2, of the remaining 4 of the denominator, against the factor 2 of 26, in the numerator, our expression, by this means, becomes, 1 cord $\times \frac{13}{1} \times \frac{3}{2} = 19\frac{1}{2}$ cords, for the answer.

4. If 11 men can cut 49 cords of wood in 7 days, when they work 14 hours per day, how many men will it take to cut 140 cords in 28 days, by working 10 hours each day?

Our expression will become,

$$\textit{Ans.}\ 11\text{ men} \times \overbrace{\tfrac{140}{49}}^{\text{Ratio of cords.}} \times \overbrace{\tfrac{7}{28}}^{\text{Ratio of days.}} \times \overbrace{\tfrac{14}{10}}^{\text{Ratio of hours.}} = 11\text{ men.}$$

5. If 100 men, in 40 days, of 10 hours each, build a wall 30 feet long, 8 feet high, and 2 feet thick, how many men must be employed to build a wall 40 feet in length, 6 feet high, and 4 feet thick, in 20 days, by working 8 hours each day?

$$\textit{Ans.}\ 100\text{ men} \times \overbrace{\tfrac{40}{20}}^{\text{Ratio of days.}} \times \overbrace{\tfrac{10}{8}}^{\text{Ratio of hours.}} \times \overbrace{\tfrac{40}{30}}^{\text{Ratio of lengths.}} \times \overbrace{\tfrac{6}{8}}^{\text{Ratio of heights.}} \times \overbrace{\tfrac{4}{2}}^{\text{Ratio of thicknesses.}} = 500\text{ men.}$$

6. If a man perform a journey of 1250 miles in 17 days, by traveling 13 hours a day, how many days will it take him to perform a journey of 1007 miles, by traveling 10 hours each day? *Ans.* $17\frac{10047}{12500}$ days.

7. If 10 cows eat 8 tons of hay in 6 weeks, how many cows will eat 56 tons in 21 weeks?

Ans. 20 cows.

8. If 8 men will mow 36 acres of grass in 9 days,

by working 9 hours each day, how many men will be required to mow 48 acres in 12 days, by working 12 hours each day? *Ans.* 6 men.

9. If 12 ounces of wool make $2\frac{1}{2}$ yards of cloth, which is 6 quarters wide, how many pounds will it take to make 150 yards of cloth, 4 quarters wide?

Ans. 30 pounds.

10. If 12 men can build a wall 26 feet long, 7 feet high, and 5 feet thick, in 20 days, in how many days will 28 men build a wall 156 feet long, 10 feet high, and 3 feet thick? *Ans.* $44\frac{4}{49}$ days.

11. If the wages of 6 men for 14 days be $84, what will be the wages of 9 men for 16 days? *Ans.* $144.

12. If a pile of wood, $30\frac{1}{2}$ feet long, 4 feet wide, and 6 feet high, is worth $25, how much is a pile 60 feet in length, 3 feet wide, and 4 feet high, worth?

Ans. $$24\frac{36}{61}$.

13. If 168 bushels of corn, when corn is worth 60 cents a bushels, be given for the carriage of 120 barrels of flour 60 miles, how many bushels of corn, when corn is worth 70 cents a bushel, must be given for the carriage of 80 barrels of flour 230 miles?

Ans. 368 bushels.

14. A wall was to be built 700 yards long in 29 days; after 12 men had been employed on it for 11 days, it was found they had built only 220 yards. How many more men must be put on it to finish it in the given time?

700 yds. = whole length of wall.

220 yds. = part built in 11 days.

480 yds. = part to be built in the remaining $29-11=18$ days. The question, therefore, may take this more

simple form: If 12 men in 11 days build 220 yards of wall, how many men will be necessary to build 480 yards in 18 days?

This, when wrought by the rule, gives

$$12 \text{ men} \times \tfrac{11}{18} \times \tfrac{480}{220} = 16 \text{ men}.$$

Now, as there are already 12 men upon the work, it becomes necessary to add 4 men more.

15. In how many days, working 9 hours a day, will 24 men dig a trench 420 yards long, 5 yards wide, and 3 yards deep, if 248 men, working 11 hours a day, in 5 days, dig a trench 230 yards long, 3 yards wide, and 2 yards deep?

Ans. $5 \text{ days} \times \tfrac{248}{24} \times \tfrac{11}{9} \times \tfrac{420}{230} \times \tfrac{5}{3} \times \tfrac{3}{2} = 288\tfrac{59}{207}$ days.

16. If 25 pears can be bought for 10 lemons, and 28 lemons for 18 pomegranates, and 1 pomegranate for 48 almonds, and 50 almonds for 70 chestnuts, and 108 chestnuts for $2\frac{1}{4}$ cents, how many pears can I buy for $1·35?

Ans. $25 \text{ pears} \times \tfrac{28}{10} \times \tfrac{1}{18} \times \tfrac{50}{48} \times \tfrac{108}{70} \times \dfrac{135}{2\frac{1}{4}} = 375$ pears.

17. Suppose that 50 men, by working 5 hours each day, can dig, in 54 days, 24 cellars, which are each 36 feet long, 21 feet wide, and 10 feet deep, how many men would be required to dig, in 27 days, 18 cellars, which are each 48 feet long, 28 feet wide, and 9 feet deep, provided they work only 3 hours each day? *Ans.* 200 men.

18. If 15 men eat 13 shillings' worth of bread in 6 days, when wheat is sold at 12 shillings per bushel, in how many days will 30 men eat $43\frac{1}{3}$ shillings' worth, when wheat is 10 shillings per bushel?

Ans. $6 \text{ days} \times \tfrac{15}{30} \times \dfrac{43\frac{1}{3}}{13} \times \tfrac{12}{10} = 12$ days.

CHAPTER VI.

SIMPLE INTEREST.

63. INTEREST is money paid by the borrower to the lender for the use of the money borrowed.

It is estimated at a certain *per cent. per annum;* that is, a certain number of dollars for the use of $100, for one year.

Thus, when $6 is paid for the use of $100, for one year, the interest is said to be at 6 *per cent.*

In the same manner, when $5 is paid for the use of $100, for one year, the interest is said to be at 5 *per cent.;* and the same for other *rates.*

The *rate per cent.* is generally fixed by law. In the New England States, the legal *rate* is 6 *per cent.*, whilst in the State of New York, it is 7 *per cent.*

The sum of money borrowed, or the sum upon which the interest is computed, is called the *principal.*

The principal, with the interest added to it, is called the *amount.*

Case I.

To find the interest on $1 for any given time at 6 *per cent.*

The interest on $100, for one year, at 6 per cent., being $6, it follows that the interest on $1, for 1 year, is $\frac{1}{100}$ of $6=$0·06; and, since 2 months is $\frac{2}{12}=\frac{1}{6}$ of one year, the interest on $1 for 2 months, is $\frac{1}{6}$ of $0·06= $0·01. Again, since 6 days is $\frac{6}{60}=\frac{1}{10}$ of 2 months, when we reckon 30 days to each month, it follows that the interest on $1, for 6 days, is $\frac{1}{10}$ of $0·01=$0·001. Hence, we have the following

RULE.

Call half the number of months, CENTS; *one-sixth the number of days,* MILLS.

EXAMPLES.

1. What is the interest of $1 for 7 months and 10 days, at 6 per cent.?

In this example, half the number of months is $3\frac{1}{2}$; which, being called cents, gives $0·035 for the interest of $1 for 7 months; again, one-sixth of the number of days is $1\frac{2}{3}$, which, being called mills, gives $0·001$\frac{2}{3}$ for the interest of $1 for 10 days; therefore, the interest for $1, for 7 months and 10 days, is $0·035+$0·001$\frac{2}{3}$= $0·036$\frac{2}{3}$.

2. What is the interest of $1 for 11 months and 11 days, at 6 per cent.? *Ans.* $0·056$\frac{5}{6}$.

3. What is the interest of $1 for 3 years, 7 months, that is, for 43 months, at 6 per cent.? *Ans.* $0·215.

4. What is the interest of $1 for 2 years, 7 months, and 9 days, at 6 per cent.? *Ans.* $0·1565.

5. What is the interest of $1 for 1 year, 7 months, and 15 days, at 6 per cent.? *Ans.* $0·0975.

6. What is the interest of $1 for 7 years and 9 days, at 6 per cent.? *Ans.* $0·4215.

Case II.

To find the interest of any given principal, for any given time, at 6 per cent., we have this

RULE.

Find the interest on $1 for the given time, by Case I., Art. **63.** *Then multiply the interest thus found by the number of dollars in the given principal.*

EXAMPLES.

1. What is the interest of $49·37 for 13 months and 15 days, at 6 per cent.?

In this example, we find the interest on $1 for 13 months and 15 days, at 6 per cent., to be $0·0675; which, multiplied by $49·37, gives $3·332475, for the interest on $49·37, for the given time.

2. What is the interest of $608·62 for 1 year and 9 months, at 6 per cent.? *Ans.* $63·9051.

3. What is the interest of $341·13 for 7 years and 9 days, at 6 per cent.? *Ans.* $143·786295.

4. What is the interest of $100 for 16 years and 8 months, at 6 per cent.? *Ans.* $100.

5. What is the interest of $591·03 for 4 years, 3 months, and 7 days, at 6 per cent.? *Ans.* $151·402185.

6. What is the interest of $0·134 for 4 months and 3 days, at 6 per cent.? *Ans.* $0·002747.

Case III.

To find the interest on any given principal, for any given time, at any given rate per cent., we have this

RULE.

Find the interest on the given principal, for the given time, at 6 *per cent., by Case II., Art.* **63.** *Then increase, or decrease, this interest by the same part of itself as it would be necessary to increase, or decrease* 6, *in order to make it agree with the given rate per cent.*

EXAMPLES.

1. What is the interest of $19·41 for 1 year, 7 months, and 13 days, at 7 per cent.?

In this example, we find, by Case II., that the interest of $19·41 for 1 year, 7 months, and 13 days, at 6 per cent., is $1·886005. Since 6, increased by its sixth part, equals 7, it will be necessary to increase the interest just found, for 6 per cent., by its sixth part, which thus becomes $2·200339$\frac{1}{6}$, for the interest at 7 per cent.

2. What is the interest of $530 for 3 years and 6 months, at 5 per cent.? *Ans.* $92·75.

In this example, it was necessary to decrease the interest at 6 per cent. by its sixth part.

3. What is the interest of $5·37 for 4 years and 12 days, at 8 per cent.? *Ans.* $1·73272.

In this example, we increased the interest at 6 per cent. by its third part.

4. What is the interest of $4070 for 3 months, at 9 per cent.? *Ans.* $91·575.

5. What is the interest of $3671 for 6 months, at 10 per cent.? *Ans.* $183·55.

6. What is the interest of $4920·05 for 3 months, at 4 per cent.? *Ans.* $49·2005.

7. What is the interest of $40·17 for 3 months and 18 days, at 3 per cent.? *Ans.* $0·36153.

8. What is the interest of $37·13 for 5 months and 12 days, at $4\frac{1}{2}$ per cent.? *Ans.* $0·7518825.

NOTE.—When the principal is given in English money, we must reduce the shillings, pence, and farthings to the decimal of a pound, and then proceed as in Federal money.

9. What is the interest of £75 13 s. 6 d. for 3 years and 5 months, at 6 per cent.?

In this example, 13 s. 6 d., reduced to the decimal of a pound, is £0·675, so that our principal is £75·675; the interest on £1 for 3 years and 5 months, at 6 per cent., is £0·205, which, multiplied into 75·675, the number of pounds, gives £15·513375=£15 10 s. $3\frac{21}{100}$ d for the interest required.

10. What is the interest of £14 5 s. $3\frac{1}{2}$ d. for 4 years, 6 months, and 14 days, at 7 per cent.?

Ans. £4 10 s. $7\frac{72}{100}$ d., nearly.

11. What is the interest of £1 7 s. 6 d. for 2 years and 6 months, at $4\frac{1}{2}$ per cent.? *Ans.* £0 3 s. $1\frac{1}{8}$ d.

12. What is the interest of £105 10 s. 6 d. for $9\frac{1}{2}$ months, at 5 per cent.? *Ans.* £4 3 s. 6 d. $1\frac{95}{100}$ far.

When the interest is 7 per cent., it may be readily computed by the assistance of the following table, which gives the interest of $1, or £1, for any time, in months and days, not exceeding 1 year.

INTEREST TABLE AT SEVEN PER CENT.

Days.	0 Month.	1 Month.	2 Months.	3 Months.	4 Months.	5 Months.
0	0·00000	0·00583	0·01167	0·01750	0·02333	0·02917
1	0·00019	0·00603	0·01186	0·01769	0·02353	0·02936
2	0·00039	0·00622	0·01206	0·01789	0·02372	0·02956
3	0·00058	0·00642	0·01225	0·01808	0·02392	0·02975
4	0·00078	0·00661	0·01244	0·01828	0·02411	0·02994
5	0·00097	0·00681	0·01264	0·01847	0·02431	0·03014
6	0·00117	0·00700	0·01283	0·01867	0·02450	0·03033
7	0·00136	0·00719	0·01303	0·01886	0·02469	0·03053
8	0·00156	0·00739	0·01322	0·01906	0·02489	0·03072
9	0·00175	0·00758	0·01342	0·01925	0·02508	0.03092
10	0·00194	0·00778	0·01361	0·01944	0·02528	0·03111
11	0·00214	0·00797	0·01381	0·01964	0·02547	0·03131
12	0·00233	0·00817	0·01400	0·01983	0·02567	0·03150
13	0·00253	0·00836	0·01419	0·02003	0·02586	0·03169
14	0·00272	0·00856	0·01439	0·02022	0·02606	0·03189
15	0·00292	0·00875	0·01458	0·02042	0·02625	0·03208
16	0·00311	0·00894	0·01478	0·02061	0·02644	0·03228
17	0·00331	0·00914	0·01497	0·02081	0·02664	0·03247
18	0·00350	0·00933	0·01517	0·02100	0·02683	0·03267
19	0·00369	0·00953	0·01536	0·02119	0·02703	0·03286
20	0·00389	0·00972	0·01556	0·02139	0·02722	0·03306
21	0·00408	0·00992	0·01575	0·02158	0·02742	0·03325
22	0·00428	0·01011	0·01594	0·02178	0·02761	0·03344
23	0·00447	0·01031	0·01614	0·02197	0·02781	0·03364
24	0·00467	0·01050	0·01633	0·02217	0·02800	0·03383
25	0·00486	0·01069	0·01653	0·02236	0·02819	0·03403
26	0·00506	0·01089	0·01672	0·02256	0·02839	0·03422
27	0·00525	0·01108	0·01692	0·02275	0·02858	0·03442
28	0·00544	0·01128	0·01711	0·02294	0·02878	0·03461
29	0·00564	0·01147	0·01731	0·02314	0·02897	0·03481
30	0·00583	0·01167	0·01750	0·02333	0·02917	0·03500

INTEREST TABLE AT SEVEN PER CENT.

Days.	6 Months.	7 Months.	8 Months.	9 Months.	10 Months.	11 Months.
0	0·03500	0·04083	0·04667	0·05250	0·05833	0·06417
1	0·03519	0·04103	0·04686	0·05269	0·05853	0·06436
2	0·03539	0·04122	0·04706	0·05289	0·05872	0·06456
3	0·03558	0·04142	0·04725	0·05308	0·05892	0·06475
4	0·03578	0·04161	0·04744	0·05228	0·05911	0·06494
5	0·03597	0·04181	0·04764	0·05347	0·05931	0·06514
6	0·03617	0·04200	0·04783	0·05367	0·05950	0·06533
7	0·03636	0·04219	0·04803	0·05386	0·05969	0·06553
8	0·03656	0·04239	0·04822	0·05406	0·05989	0·06572
9	0·03675	0·04258	0·04842	0·05425	0·06008	0·06592
10	0·03694	0·04278	0·04861	0·05444	0·06028	0·06611
11	0·03714	0·04297	0·04881	0·05464	0·06047	0·06631
12	0·03733	0·04317	0·04900	0·05483	0·06067	0·06650
13	0·03753	0·04336	0·04919	0·05503	0·06086	0·06669
14	0·03772	0·04356	0·04939	0·05522	0·06106	0·06689
15	0·03792	0·04375	0·04958	0·05542	0·06125	0·06708
16	0·03811	0·04394	0·04978	0·05561	0·06144	0·06728
17	0·03831	0·04414	0·04997	0·05581	0·06164	0·06747
18	0·03850	0·04433	0·05017	0·05600	0·06183	0·06767
19	0·03869	0·04453	0·05036	0·05619	0·06203	0·06786
20	0·03889	0·04472	0·05056	0·05639	0·06222	0·06806
21	0·03908	0·04492	0·05075	0·05658	0·06242	0·06825
22	0·03928	0·04511	0·05094	0·05678	0·06261	0·06844
23	0·03947	0·04531	0·05114	0·05697	0·06281	0·06864
24	0·03967	0·04550	0·05133	0·05717	0·06300	0·06883
25	0·03986	0·04569	0·05153	0·05736	0·06319	0·06903
26	0·04006	0·04589	0·05172	0·05756	0·06339	0·06922
27	0·04025	0·04608	0·05192	0·05775	0·06358	0·06942
28	0·04044	0·04628	0·05211	0·05794	0·06378	0·06961
29	0·04064	0·04647	0·05231	0·05814	0·06397	0·06981
30	0·04083	0·04667	9·05250	0·05833	0·06417	0·07000

We will now give, to be wrought by the aid of the preceding table, the following

EXAMPLES.

1. What is the interest of \$37·13 for 3 months and 3 days, at 7 per cent.?

Operation.

\$0·01808 - -	Tabular number, 3 mo. 3 days.
37·13 - -	Multiply by the number of dollars in the principal.
5424	
1808	
12656	
5424	
\$0·6713104,	Interest sought.

2. What is the interest of \$320 for 6 months and 3 days, at 7 per cent.? *Ans.* \$11·386, nearly.

3. What is the interest of £20 5 s. 6 d. for 8 months 17 days, at 7 per cent.?

Reducing the shillings and pence to the decimal of a pound, our principal will become £20·275. Hence, multiplying £0·04997, the tabular number for 8 months 17 days, by 20·275, we find £1·01314175. Reducing this decimal of a pound to shillings and pence, we have £1 0 s. 3 d., nearly for the interest required.

4. What is the interest of \$500 for 6 months and 1 day, at 7 per cent.? *Ans.* \$17·595.

5. What is the interest of £500 10 s. for 4 months 15 days, at 7 per cent.? *Ans.* £13 2 s. $9\frac{15}{100}$ d.

6. What is the interest of \$1250 for 3 years and 3 months, at 7 per cent.?

The tabular number for 3 months is $0·01750; to which, adding $0·21, the interest of $1 for 3 years, we have $0·2275, for the interest of $1 for 3 years and 3 months. Now, multiplying $0·2275 by 1250, we obtain $284·375, for the interest sought.

7. What is the interest of $33·33 for 2 years, 5 months, and 3 days, at 7 per cent.? *Ans.* $5·658, nearly.

PARTIAL PAYMENTS.

64. When notes, bonds, or obligations, receive partial payments, or indorsements, we must use the following Rule, which was given by Chancellor Kent, in the *New York Chancery Reports:*

RULE.

"*The Rule for casting interest, when partial payments have been made, is to apply the payment in the first place to the discharge of the interest then due. If the payment exceed the interest, the surplus goes towards discharging the principal, and the subsequent interest is to be computed on the balance of the principal remaining due. If the payment be less than the interest, the surplus of interest must not be taken to augment the principal; but interest continues on the former principal until the period when the payments, taken together, equal or exceed the interest due, and then the surplus is to be applied towards discharging the principal, and interest is to be computed on the balance, as aforesaid.*"

UTICA, *May* 16, 1839.

1. For value received, I promise to pay A. B., or order, three hundred and twenty dollars, with interest at 7 per cent. C. D.

Indorsements were made on this note as follows:

September 3, 1839, there was paid	. . .	$30·00
January 5, 1840, " " "	. . . . .	5·00
August 11, 1841, " " "	. . .	40·00
February 25, 1842, " " "	. . .	100·00

How much was due May 3, 1843?

We first find the successive periods of time for which interest is to be computed by the following

Operation.

	Year.	*Mo.*	*Day.*	*Year.*	*Mo.*	*Day.*
Date of note,	1839	4	16			
1st indorsement,	1839	8	3		3	17
2d indorsement,	1840	0	5		4	2
3d indorsement,	1841	7	11	1	7	6
4th indorsement,	1842	1	25		6	14
Date of settlement,	1843	4	3	1	2	8

The interest being 7 per cent., we will compute it by the aid of the table under Art. **63.**

Amount of the note, or the principal, is . .	$320·000
Interest up to Sept. 3, 1839, is	6·659
	326·659
Deduct first indorsement, ...	30·000
	296·659

Interest up to January 5, 1840, is $7·037, which is greater than the second indorsement.

Brought forward, ...	$296·659
Interest from Sept. 3, 1839, to August 11, 1841, is	40·263
	336·922
Sum of second and third indorsements, . . .	45·000
	291·922
Interest up to February 25, 1842, is	11·011
	302·933
Deduct fourth indorsement, ...	100·000
	202·933
Interest up to May 3, 1843, is	16·888
Amount due May 3, 1843, is	$219·821

UTICA, *Nov.* 1, 1837.

2. For value received, I promise to pay THOMAS JONES, or order, the sum of six hundred and twenty dollars, on demand, with interest.

CHARLES BANK.

The following indorsements were made on this note:

1838, Oct. 6,	there was indorsed . . .	$61·07
1839, March 4,	" " " . . .	89·03
1839, Dec. 11,	" " " . . .	107·77
1840, July 20,	" " " . . .	200·50

What was the balance due Oct. 15, 1840, allowing 7 per cent. interest?

	Year.	*Mo.*	*Day.*	*Mo.*	*Day.*
Date of note,	1837	10	1		
1st indorsement,	1838	9	6	11	5
2d indorsement,	1839	2	4	4	28
3d indorsement,	1839	11	11	9	7
4th indorsement,	1840	6	20	7	9
Date of settlement,	1840	9	15	2	25

The amount of note, or principal, is . . .	$620·000
Interest on the same, up to Oct. 6, 1838, is	40·386
Amount due on note, Oct. 6, 1838, is . . .	660·386
The first indorsement is	61·070
	599·316
Interest from Oct. 6, 1838, to March 4, 1839, is	17·247
Amount due March 4, 1839, is	616·563
The second indorsement is	89·030
	527·533
Interest from March 4, 1839, to Dec. 11, 1839, is	28·414
	555·947
The third indorsement is	107·770
	448·177
Interest from Dec. 11, 1839, to July 20, 1840, is	19·085
	467·262
The fourth indorsement is	200·500
	266·762
Interest from July 20, 1840, to Oct. 15, 1840, is	4·409
Ans	$271·171

UTICA, *May* 1, 1836.

3. For value received, I promise to pay ISAAC CLARK, or order, three hundred and forty-nine dollars, ninety-nine cents, and eight mills, with interest at 6 per cent.

N. BROWN.

Indorsements were made on this note as follows:

Dec. 25, 1836, there was paid	$49·998
June 30, 1837, " " "	4·998
Aug. 22, 1838, " " "	15·000
June 4, 1839, " " "	99·999

How much was due April 5, 1840?

	Year.	Mo.	Day.	Year.	Mo.	Day.	Interest at 6 per cent.
Date of note	1836	4	1				
1st indorsement,	1836	11	25		7	24	$0·039
2d "	1837	5	30		6	5	$0·030\frac{5}{6}$
				1	1	22	$0·068\frac{2}{3}$
3d "	1838	7	22				
					9	12	0·047
4th "	1839	5	4		10	1	$0·050\frac{1}{6}$
Date of settlement,	1840	3	5				

The amount of the note, or principal, is . .		$349·998
Interest up to Dec. 25, 1836, is		13·650
		363·648
The first indorsement is		49·998
		313·650
Interest up to June 4, 1839, is		45·950
		359·600
Indorsement June 30, 1837, which is less than the interest then due,	$4·998	
Indorsement August 22, 1838, . . .	15·000	
	19·998	
This sum is still less than the interest now due.		
Indorsement June 4, 1839, ...	99·999	
		$119·997
This sum exceeds the interest now due.		
		239·603
Interest up to April 5, 1840, is		12·020
Amount due April 5, 1840, ...		$251·623

UTICA, *Dec.* 9, 1835.

4. For value received, I promise to pay PETER SMITH, or order, one hundred and eight dollars and forty-three cents, on demand, with interest at 7 per cent.

JOHN SAVEALL.

Indorsements were made as follows :

March 3, 1836, there was indorsed . . . $50·04
Dec. 10, 1836, " " " . . . 13·19
May 1, 1838, " " " . . . 50·11

How much remained due Oct. 9, 1840?

Ans. $5·844.

UTICA, *Aug.* 1, 1837.

5. For value received, I promise to pay F. GOULD, or bearer, one hundred and forty-three dollars and fifty cents, on demand, with interest

D. FARLING.

Dec. 17, 1837, there was indorsed . . . $37·40
July 1, 1838, " " " . . . . . 7·09
Dec. 22, 1839, " " " . . . 13·13
Sept. 9, 1840, " " " . . . . 50·50

How much remains due Dec. 28, 1840, the interest being 7 per cent.? *Ans.* $60·866.

6. A note of $486 is dated Sept. 7, 1831, on which,
March 22, 1832, there was paid $125
Nov. 29, 1832, " " " 150
May 13, 1833, " " " 120

What was the balance due April 19, 1834, the interest being 7 per cent.? *Ans.* $144·404.

---

65. The principal, the rate per cent., the time, and the interest, are so related to each other, that any three of them being given, the remaining one can be found.

Problem I.

Given the principal, the rate per cent., and the time, to find the interest. The rule for this problem has already been given under Case III., Art. **63**; it is equivalent to the following

RULE.

Multiply the interest of $1 *for the given time and given rate per cent., by the number of dollars in the principal.*

Problem II.

Given the time, the rate per cent., and the interest, to find the principal. By the reverse of the last problem, we obtain this

RULE.

Divide the given interest by the interest of $1 *for the given time and given rate per cent.; and the quotient will be the number of dollars in the principal.*

EXAMPLES.

1. The interest on a certain principal for 9 months and 10 days, at $4\frac{1}{2}$ per cent., is $1·01605. What was the principal?

In this example, we find the interest of $1 for 9 months and 10 days, at $4\frac{1}{2}$ per cent., to be $0·035; ∴ $1·01605, divided by $0·035, gives 29·03 for the number of dollars in the principal required.

2. What principal will, in 1 year, 7 months, and 15 days, at 6 per cent., give $9·75 interest? *Ans.* $100.

3. What principal will, in 7 years and 9 days, at 6 per cent., give $16·86 interest? *Ans.* $40.

4. What principal will, in 3 years and 6 months, at 5 per cent., give $92·75 interest? *Ans.* $530.

5. What principal will, in 3 months and 9 days, at 8 per cent., give $90, interest? *Ans.* $4090·909.

Problem III.

Given the principal, the time, and the interest, to find the rate per cent.

RULE.

Divide the given interest by the interest of the given principal, for the given time, at one per cent.

EXAMPLES.

1. The interest of $100 for 9 months and 10 days, is $3·50. What is the rate per cent.?

In this example, we find the interest of $100 for 9 months and 10 days, at 6 per cent., to be $4·66$\frac{2}{3}$. The interest at 1 per cent. is $\frac{1}{6}$ of $4·66$\frac{2}{3}$=$0·77$\frac{7}{9}$; therefore, dividing $3·50 by $0·77$\frac{7}{9}$, we obtain 4$\frac{1}{2}$ for the rate per cent. required.

2. At what rate per cent. will $530, in 3 years and 6 months, give $92·75, interest? *Ans.* 5 per cent.

3. At what rate per cent. will $19·41, in 1 year, 7 months, and 13 days, give $2·200339$\frac{1}{6}$, interest?

Ans. 7 per cent.

4. At what rate per cent. will $5·37, in 4 years and 12 days, give $1·73272, interest? *Ans.* 8 per cent.

5. At what rate per cent. will \$4070, in 3 months, give \$91·575, interest? *Ans.* 9 per cent.

Problem IV.

Given the principal, the rate per cent., and the interest, to find the time.

RULE.

Divide the given interest by the interest of the given principal for 1 *year at the given rate per cent.*

EXAMPLES.

1. In what time will \$37·13, at $4\frac{1}{2}$ per cent., give \$0·7518825, interest?

In this example, we find the interest of \$37·13 for 1 year, at $4\frac{1}{2}$ per cent., to be \$1·67085; ∴ dividing \$0·7518825 by \$1·67085, we get 0·45, which, considered as years, gives, when reduced, 5 months and 12 days.

2. In what time will \$700, at 7 per cent., give \$85·75, interest? *Ans.* 1 year, 9 months.

3. In what time will \$100, at 6 per cent., give \$100 interest? That is, in what time will a given principal double itself at 6 per cent. interest? *Ans.* $16\frac{2}{3}$ years.

4. In what time will a given principal double itself at 7 per cent.? *Ans.* $14\frac{2}{7}$ years.

5. In what time will a given principal double itself at 8 per cent.? *Ans.* $12\frac{1}{2}$ years.

6. In what time will a given principal double itself at 5 per cent.? *Ans.* 20 years.

7. In what time will a given principal double itself at $4\frac{1}{2}$ per cent.? *Ans.* $22\frac{2}{9}$ years.

To find the number of years required for a given sum to double itself at simple interest, we have only to divide 100 by the rate per cent.

The following table gives the time required for a given principal to double itself at simple interest.

Per cent.	Years.	Per cent.	Years.	Per cent.	Years.
1	100	4	25	7	$14\frac{2}{7}$
$1\frac{1}{2}$	$66\frac{2}{3}$	$4\frac{1}{2}$	$22\frac{2}{9}$	$7\frac{1}{2}$	$13\frac{1}{3}$
2	50	5	20	8	$12\frac{1}{2}$
$2\frac{1}{2}$	40	$5\frac{1}{2}$	$18\frac{2}{11}$	$8\frac{1}{2}$	$11\frac{13}{17}$
3	$33\frac{1}{3}$	6	$16\frac{2}{3}$	9	$11\frac{1}{9}$
$3\frac{1}{2}$	$28\frac{4}{7}$	$6\frac{1}{2}$	$15\frac{5}{13}$	$9\frac{1}{2}$	$10\frac{10}{19}$

DISCOUNT.

66. Discount is an allowance made for the payment of money before it is due.

The *present worth* of a debt payable at some future time, without interest, is such a sum of money as will, in the given time, amount to the debt.

When the interest is at 6 per cent., the amount of $1 for one year, is $1·06; therefore, the present worth of $1·06, due one year hence, is $1. We may also infer that the present worth of any sum for one year will be as many dollars as $1·06 is contained in the given sum. Hence, we have the following

RULE.

Find the amount of $1 *for the given time, at the given rate per cent.; then divide the sum by this*

amount, and it will give the number of dollars in the present worth.

Subtract the present worth from the amount, and it will give the discount.

EXAMPLES.

1. What is the present worth of \$622·75, due 3 years and 6 months, at 5 per cent.?

In this example, we find the amount of \$1 for 3 years and 6 months, at 5 per cent., to be \$1·175; therefore, dividing \$622·75 by \$1·175, we get 530 for the number of dollars in the present worth. If we subtract the present worth from the sum, we get \$92·75 for the discount.

2. What is the present worth of \$4161·575, due 3 months hence, at 9 per cent.? *Ans.* \$4070.

3. What is the present worth of \$7·10272, due 4 years and 12 days hence, at 8 per cent.? *Ans.* \$5·37.

4. Sold goods for \$1500, to be paid one half in 6 months, and the other half in 9 months; what is the present worth of the goods, interest being at 7 per cent.? *Ans.* \$1437·227.

5. What is the present worth of \$50, payable at the end of 3 months, at 7 per cent.? *Ans.* \$49·14.

6. What is the discount on \$100, due 6 months hence, at 6 per cent.? *Ans.* \$2·91̇3.

The following table gives the present worth of \$1, or £1, for months and days, for any time not exceeding 1 year, at 7 per cent.

TABLE OF PRESENT WORTHS.

Days.	0 Month.	1 Month.	2 Months.	3 Months.
0	0·000000	0·994200	0·988468	0·982801
1	0·999806	0·994008	0·988278	0·982613
2	0·999600	0·993816	0·988088	0·982425
3	0·999417	0·993624	0·987898	0·982238
4	0·999223	0·993432	0·987709	0·982050
5	0·999029	0·993240	0·987519	0·981863
6	0·998835	0·993049	0·987329	0·981675
7	0·998641	0·992857	0·987140	0·981488
8	0·998447	0·992665	0·986950	0·981301
9	0·998253	0.992474	0·986761	0·981114
10	0·998059	0·992282	0·986572	0·980926
11	0·997866	0·992091	0·986382	0·980739
12	0·997672	0·991899	0·986193	0·980552
13	0·997479	0·991708	0·986004	0·980365
14	0·997285	0·991517	0·985815	0·980179
15	0·997092	0·991326	0·985626	0·979992
16	0·996899	0·991135	0·985437	0·979805
17	0·996705	0·990944	0·985249	0·979618
18	0·996512	0·990753	0·985060	0·979432
19	0·996319	0·990562	0·984871	0·979245
20	0·996126	0·990371	0·984683	0·979059
21	0·995933	0·990181	0·984494	0·978873
22	0·995740	0·989990	0·984306	0·978686
23	0·995548	0·989800	0·984117	0·978500
24	0·995355	0·989609	0·983929	0·978314
25	0·995162	0·989419	0·983741	0·978128
26	0·994970	0·989228	0·983553	0·977942
27	0·994777	0·989038	0·983365	0·977756
28	0·994585	0·988848	0·983177	0·977570
29	0·994393	0·988658	0·982989	0·977384
30	0·994200	0·988468	0·982801	0·977199

TABLE OF PRESENT WORTHS.

Days.	4 Months.	5 Months.	6 Months.	7 Months.
0	0·977199	0·971660	0·966184	0·960769
1	0·977013	0·971476	0·966002	0·960589
2	0·976827	0·971293	0·965821	0·960410
3	0·976642	0·971109	0·965639	0·960230
4	0·976457	0·970926	0·965458	0·960051
5	0·976271	0·970743	0·965277	0·959872
6	0·976086	0·970560	0·965096	0·959693
7	0·975901	0·970377	0·964915	0·959514
8	0·975716	0·970193	0·964734	0·959335
9	0·975530	0·970011	0·964553	0·959156
10	0·975345	0·969828	0·964372	0·958977
11	0·975160	0·969645	0·964191	0·958798
12	0·974976	0·969462	0·964010	0·958620
13	0·974791	0·969279	0·963830	0·958441
14	0·974606	0·969097	0·963649	0·958262
15	0·974421	0·968914	0·963468	0·958084
16	0·974237	0·968731	0·963288	0·957905
17	0·974052	0·968549	0·963108	0·957727
18	0·973868	0·968367	0·962927	0·957549
19	0·973683	0·968184	0·962747	0·957370
20	0·973499	0·968002	0·962567	0·957192
21	0·973315	0·967820	0·962387	0·957014
22	0·973131	0·967638	0·962207	0·956836
23	0·972947	0·967456	0·962027	0·956658
24	0·972763	0·967274	0·961847	0·956480
25	0·972579	0·967092	0·961667	0·956302
26	0·972395	0·966910	0·961487	0·956125
27	0·972211	0·966728	0·961307	0·955947
28	0·972027	0·966547	0·961128	0·955769
29	0·971844	0·966365	0·960948	0·955592
30	0·971660	0·966184	0·960769	0·955414

TABLE OF PRESENT WORTHS.

Days.	8 Months.	9 Months.	10 Months.	11 Months.
0	0·955414	0·950119	0·944882	0·939702
1	0·955237	0·949943	0·944708	0·939531
2	0·955059	0·949768	0·944535	0·939359
3	0·954882	0·949592	0·944361	0·939188
4	0·954705	0·949417	0·944188	0·939016
5	0·954527	0·949242	0·944015	0·938845
6	0·954350	0·949067	0·943841	0·938673
7	0·954173	0·948892	0·943668	0·938502
8	0·953996	0·948717	0·943495	0·938331
9	0·953819	0·948542	0·943322	0·938160
10	0·953642	0·948367	0·943149	0·937989
11	0·953466	0·948192	0·942976	0·937817
12	0·953289	0·948017	0·942803	0·937647
13	0·953112	0·947842	0·942630	0·937476
14	0·952936	0·947668	0·942458	0·937305
15	0·952759	0·947493	0·942285	0·937134
16	0·952583	0·947319	0·942112	0·936963
17	0·952406	0·947144	0·941940	0·936793
18	0·952230	0·946970	0·941767	0·936622
19	0·952054	0·946795	0·941595	0·936451
20	0·951877	0·946621	0·941423	0·936281
21	0·951701	0·946447	0·941250	0·936110
22	0·951525	0·946273	0·941078	0·935940
23	0·951349	0·946099	0·940906	0·935770
24	0·951173	0·945925	0·940734	0·935600
25	0·950997	0·945751	0·940562	0·935429
26	0·950821	0·945577	0·940390	0·935259
27	0·950646	0·945403	0·940218	0·935089
28	0·950470	0·945229	0·940046	0·934919
29	0·950294	0·945056	0·939874	0·934749
30	0·950119	0·944882	0·939702	0·934579

The following examples may be worked by the aid of the foregoing table :

7. What is the present worth, interest being 7 per cent., of $320, due at the end of 6 months and 3 days?

Operation.

	$0·965639 - - -	Tabular number for 6 months 3 days.
	320 - - -	Multiply by the number of dollars.
	19312780	
	2896917	
Ans.	$309·004480	

8. What is the discount on $750, due 9 months hence, at 7 per cent.? *Ans.* $37·411.

9. What is the present worth of $3471·20, due 3 months and 9 days hence, at 7 per cent.?

Ans. $3405·643.

10. What is the discount of $150, due 3 months and 18 days hence, at 7 per cent.? *Ans.* $3·085.

11. What is the discount of $961·13, due 10 months and 5 days hence, at 7 per cent.? *Ans.* $53·809.

EQUATION OF PAYMENTS,

DEDUCED BY CONSIDERING THE PRESENT VALUES OF THE SEVERAL PAYMENTS.

67. Suppose A owes me $100, due at the end of 3 months, and $100, due at the end of 9 months, and I wish him to give me one note of $200, of such a time that its present value shall be the same as the sum of the present values of the two individual debts. How long after date must this note be made payable?

By the foregoing table for the present worth, we find the present value of $100 for three months, to be $98·2801: the present value of $100 for 9 months, to be $95·0119. Taking the sum, we have $193·292 for the present value of $200, due at a future time, which time we are required to find.

We may obviously consider $193·292 as a principal, which, at the given rate per cent., will amount to $200, in the time sought. The interest is, therefore, $6·708.

This question now is equivalent to the following:

Given the principal, the rate per cent., and the interest, to find the time.

This has been solved under Prob. IV., Art. **65.** Proceeding according to this rule, we find the interest of $193·292 for 1 year, at 7 per cent., to be $13·53, nearly. Dividing $6·708 by $13·53, we get 0·4958 of a year, equal to 5 months, 28·49 days, nearly. The equated time, when found by the ordinary method, is 6 months.

From the above, we deduce the following rule for the equation of payments, founded on the principle of *equivalent present values.*

RULE.

Find the sum of the present values of the individual debts; also, the sum of their discounts. Then regard the sum of the present values of the individual debts as a principal, and the sum of their discounts as the interest. Then proceed with this principal and interest and given rate per cent., according to Rule under Prob. IV., Art. **65.**

EXAMPLE.

Suppose A owes me \$500, due at the end of 3 months, \$600 at the end of 4 months, and \$800 at the end of 6 months. How long may the whole \$1900 remain unpaid, so that its present worth may be the same as the sum of the present worths of the individual debts?

Present value of \$500, due 3 months hence, is	\$191·4005
" " 600, " 4 " "	586·3194
" " 800, " 6 " "	772·9472
Sum of their present values is	\$1850·6671

Sum of several payments	\$1900·0000
	1850·6671
Sum of their discounts is	\$49·3329

Now, if a principal of \$1850·6671 gives \$49·3329 interest at 7 per cent., what is the number of years?

By rule under Prob. IV., we find that the interest on \$1850·6671 for 1 year, at 7 per cent., to be \$129·5467. Hence, dividing \$49·3329 by \$129·5467, we find 0·3808 of a year, which is the same as 4 months, 17·088 days, nearly.

If we find the equated time by the usual rule for equation of payments, it will be 4 months, 17·368 days, nearly, which differs less than half a day from the result by the above method.

Other examples might be given to illustrate this method of finding the time for the equation of payments, on the principle of *equivalent present values*, but enough has been done to call the attention of the student to this singular subject.

CHAPTER VII.

COMPOUND INTEREST.

68. WHEN, at the end of each year, the interest due is added to the principal, and the amount thus obtained is considered as a new principal, upon which the interest is cast for another year, and added to it to form a new principal for the next year, and so on to the last year, the last amount thus obtained, is called the AMOUNT AT COMPOUND INTEREST. If from this amount we subtract the original principal, we obtain the COMPOUND INTEREST.

EXAMPLES.

1. What is the compound interest of $1000 for 3 years, at 7 per cent.?

Principal, ...	$1000
Interest on $1000 for one year, ...	70
First amount, or second principal, ...	1070
Interest on $1070 for one year, ...	74·90
Second amount, or third principal, ...	1144·90
Interest on $1144·90 for one year, ...	80·143
Third amount, ...	1225·043
Original principal, ...	1000
The compound interest required, . *Ans.*	$225·043

2. What is the compound interest of $100 for 4 years, at 6 per cent.?

Principal, ...	$100
Interest for first year	6
First amount, or second principal	106
Interest for second year, ...	6·36
Second amount, or third principal, ...	112·36
Interest for third year, ...	6·74
Third amount, or fourth principal, ...	119·10
Interest for fourth year, ...	7·15
Fourth amount, ...	126·25
Original principal, ...	100
Compound interest required, . . . *Ans.*	$26·25

3. What is the compound interest of $630 for 4 years, at 5 per cent.? *Ans.* $135·769.

By carefully reviewing the above manner in which compound interest is computed, we discover that the successive amounts, which are considered as new principals, form the terms of a geometrical series, whose first term is the original principal; the ratio is the amount of $1 for one year, at the given rate per cent.; the number of terms is equal to the number of years, *plus* one.

From this we learn that finding the amount of a given principal, for a given number of years, at a given rate per cent., consists in finding the last term of a geometrical progression, when the first term, the ratio, and the number of terms are given.

Thus, the amount of $1 for one year, at 3 per cent., is $1·03; for two years, it is $$(1·03)^2$; for three years, it is $$(1·03)^3$; for four years, it is $$(1·03)^4$; and in general, for any number of years, it is found by raising 1·03 to a power denoted by the number of years.

TABLE,

Showing the amount of **$1**, or **£1**, for any number of years, not exceeding 30, at 3, 4, 5, and 6 per cent., at compound interest, the interest being compounded *yearly*.

Years.	3 per cent.	4 per cent.	5 per cent.	6 per cent.
1	1·030000	1·040000	1·050000	1·060000
2	1·060900	1·081600	1·102500	1·123600
3	1·092727	1·124864	1·157625	1·191016
4	1·125509	1·169859	1·215506	1·262477
5	1·159274	1·216653	1·276282	1·338225
6	1·194052	1·265319	1·340096	1·418519
7	1·229874	1·315932	1·407100	1·503630
8	1·266770	1·368569	1·477455	1·593848
9	1·304773	1.423312	1·551328	1·689479
10	1·343916	1·480244	1·628895	1·790848
11	1·384234	1·539454	1·710339	1·898299
12	1·425761	1·601032	4·795856	2·012196
13	1·468534	1·665074	1·885649	2·132928
14	1·512590	1·731676	1·979932	2·260904
15	1·557968	1·800944	2·078928	2·396558
16	1·604707	1·872981	2·182875	2·540352
17	1·652848	1·947900	2·292018	2·692773
18	1·702433	2·025817	2·406619	2·854339
19	1·753506	2·106849	2·526950	3·021599
20	1·806111	2·191123	2·653298	3·207135
21	1·860295	2·278768	2·785963	3·399564
22	1·916103	2·369919	2·925261	3·603537
23	1·973587	2·464716	3·071524	3·819750
24	2·032794	2·563304	3·225100	4·048935
25	2·093778	2·665836	3·386355	4·291871
26	2·156591	2·772470	3·555673	4·549383
27	2·221289	2·883369	3·733456	4·822346
28	2·287928	2·998703	3·920129	5·111637
29	2·356566	3·118651	4·116136	5·418388
30	2·427262	3·243398	4·321942	5·743491

We will now solve the following questions by means of the preceding table.

4. What is the amount of $790 for 13 years, at 6 per cent.?

From our table, we find the amount of $1 for 13 years, at 6 per cent., to be $2·132928; this, multiplied by 790, the number of dollars in the principal, gives $1685·013 for the amount required.

5. What is the compound interest of $49, for 20 years, at 5 per cent.?

In this example, we find, from the table, that the amount of $1 for 20 years, at 5 per cent., is $2·653298; which, multiplied by 49, gives $130·012 for the amount of $49, from which, if we subtract $49, we get $81·012 for the compound interest required.

6. What is the compound interest of $100 for 17 years, at 6 per cent.? *Ans.* $169·277.

7. What is the compound interest of $375 for 20 years, at 6 per cent.? *Ans.* $827·676.

8. What is the amount of $875 for 12 years, at 6 per cent., compound interest? *Ans.* $1760·672.

9. What is the amount of $625 for 18 years, at 5 per cent., compound interest? *Ans.* $1504·137.

10. What is the amount of $379 for 30 years, at 3 per cent., compound interest? *Ans.* $919·932.

Note.—When the interest is compounded *half-yearly*, we must take the amount of $1 for half a year, and raise it to a power denoted by the number of half-years in the whole time; this power, multiplied by the principal, will give the amount. We must proceed in a similar way for any other aliquot part of a year. Or, in such cases, we may make use of our table, as in the work of next question.

11. What is the amount of $100 for 3 years, at 6 per

cent. per annum, when the interest is added at the end of every 6 months?

In this example, we change the 6 per cent. to 3 per cent., and the 3 years to 6 years; we then find the tabular number to be $1·194052; which, multiplied by 100, gives $119·405 for the amount required.

12. What will £600 amount to in 6 years, at 8 per cent., compound interest, supposing the interest to be receivable half-yearly? *Ans.* £960 12 s. 4½ d.

13. What will $890 amount to in 5 years and 4 months, at 9 per cent. per annum, compound interest, the interest being added at the end of every 4 months? *Ans.* $1428·189.

14. What will $3705 amount to in 3 years and 3 months, at 12 per cent. per annum, compound interest, the interest being added at the end of every 3 months? *Ans.* $5440·918.

15. What will $378 amount to in 7 years and 6 months, at 8 per cent. per annum, the interest being compounded half-yearly? *Ans.* $680·757.

16. What will $1000 amount to in 15 years, at 8 per cent. per annum, the interest being compounded half-yearly? *Ans.* $3243·398.

COMPOUND DISCOUNT.

69. Compound Discount is an allowance made for the payment of money before it is due, on the supposition that the money draws *compound interest.*

The *present worth* of a debt payable at some future period, without interest, is such a sum as being put out at compound interest, will, in the given time, at the given rate per cent., amount to the debt.

Hence, the finding the present worth resolves itself into the following:

Given the amount at compound interest, the time, and the rate per cent., to find the principal.

Under compound interest, it was shown that the amount was equal to the number of dollars in the principal, multiplied by the amount of $1 for one year, raised to a power whose exponent is the number of years. Hence, we have the following rule to find the principal, or present worth:

RULE.

Divide the given amount by the amount of \$1 *for* 1 *year, raised to a power whose exponent is equal to the number of years.*

EXAMPLES.

1. What is the present worth of \$1685, due 13 years hence, allowing discount according to 6 per cent., compound interest?

From the table under Art. **68**, we find that the amount of \$1 for 1 year, at 6 per cent., raised to the 13th power, or, what is the same, the number for 13 years is \$2·132928; ∴ dividing \$1685 by 2·132928, gives \$789·994 for the present worth required.

The present worth of \$1 for one year, at 3 per cent., is $\frac{1}{1{\cdot}03}$; for two years, it is $(\frac{1}{1{\cdot}03})^2$; for three years, it is $(\frac{1}{1{\cdot}03})^3$; for four years, it is $(\frac{1}{1{\cdot}03})^4$; and in general, the present worth of \$1, at compound interest, is the reciprocal of the amount of \$1, at compound interest, for the same time and same rate per cent.

TABLE;

SHOWING the PRESENT WORTH of $1, or £1, for any number of years, from 1 to 30, at 3, 4, 5, and 6 per cent., *compound discount.*

Years.	3 per cent.	4 per cent.	5 per cent.	6 per cent.
1	0·970874	0·961538	0·952381	0·943396
2	0·942596	0·924556	0·907029	0·889996
3	0·915142	0·888996	0·863838	0·839619
4	0·888487	0·854804	0·822702	0·792094
5	0·862609	0·821927	0·783526	0·747258
6	0·837484	0·790315	0·746215	0·704961
7	0·813092	0·759918	0·710681	0·665057
8	0·789409	0·730690	0·676839	0·627412
9	0·766417	0·702587	0·644609	0·591898
10	0·744094	0·675564	0·613913	0·558395
11	0·722421	0·649581	0·584679	0·526788
12	0·701380	0·624597	0·556837	0·496969
13	0·680958	0·600574	0·530321	0·468840
14	0·661118	0·577475	0·505068	0·442301
15	0·641862	0·555264	0·481017	0·417265
16	0·623167	0·533908	0·458112	0·393646
17	0·605016	0·513373	0·436297	0·371364
18	0·587395	0·493628	0·415521	0·350344
19	0·570286	0·474642	0·395734	0·330513
20	0·553676	0·456387	0·376889	0·311805
21	0·537549	0·438834	0·358942	0·294155
22	0·521892	0·421955	0·341850	0·277505
23	0·506692	0·405726	0·325571	0·261797
24	0·491934	0·390121	0·310068	0·246979
25	0·477606	0·375117	0·295303	0·232999
26	0·463695	0·360689	0·281241	0·219810
27	0·450189	0·346817	0·267848	0·207368
28	0·437077	0·333477	0·255094	0·195630
29	0·424346	0·320651	0·242946	0·184557
30	0·411987	0·308319	0·231377	0·174110

The *present worth* of a given sum of money, discounting at compound interest, is easily obtained by the preceding table.

2. How much money must be placed out at compound interest to amount to $1000 in 20 years, the interest being 5 per cent.? *Ans.* $376·889.

3. What is the present worth of $1000, due 27 years hence, discounting at 3 per cent., compound interest?

From the preceding table, we find the present worth of $1 for 27 years, at 3 per cent., to be $0·450189; this, multiplied by 1000, gives $450·189 for the present worth required.

4. What is the present worth of $3525, due in 3 years, discounting at 6 per cent., compound interest?
Ans. $2959·657.

5. What is the present worth of $350, due 5 years hence, discounting at 6 per cent., compound interest?
Ans. $261·54.

6. What is the present worth of $375, due 17 years hence, discounting at 4 per cent., compound interest?
Ans. $192·515.

7. What is the present worth of $672, due 13 years hence, discounting at 5 per cent., compound interest?
Ans. $356·376.

8. What is the present worth of $400, due 19 years hence, discounting at 6 per cent., compound interest?
Ans. $132·205.

9. What is the present worth of $111, due 29 years hence, discounting at 3 per cent., compound interest?
Ans. $47·102.

ANNUITIES.

70. AN ANNUITY is a fixed sum of money, which is paid periodically for a certain length of time.

Case I.

To find the amount of an annuity which has been forborne for a given time.

It is obvious that the last year's payment will be simply the annuity without any interest; the last but one will be the amount of the annuity for one year; the last but two will be the amount of the annuity for two years, and so on; and the sum of all these partial amounts will give the total amount due. Now we discover that these partial amounts, or payments, form a geometrical progression, whose first term is the annuity, the ratio is the amount of $1 for 1 year, and the number of terms is equal to the number of years; therefore, the amount of an annuity is found by summing the terms of a geometrical progression, when the first term, the number of terms, and the ratio, are given. This may be done by the following

RULE.

From the amount of $1 *for* 1 *year, raised to a power whose exponent is equal to the number of years, subtract* $1; *divide the remainder by the interest of* $1 *for* 1 *year; then multiply the annuity by this quotient.*

NOTE.—The different powers of the amount of $1 for one year, may be taken from the table under Art. 68.

EXAMPLES.

1. What is the amount of an annuity of $200, which has been forborne 14 years, at 6 per cent., compound interest?

From table under Art. **68**, we find the 14th power of the amount of $1 for one year, at 6 per cent., to be $2·260904; subtracting $1, and dividing the remainder by $0·06, the interest of $1 for one year, we get 21·01506; then multiplying $200, the annuity, by 21·01506, we find $4203·012 for the amount required.

2. Suppose a person, who has a salary of $700 a year, payable quarterly, to allow it to remain unpaid for 4 years; how much would be due him, allowing quarterly compound interest, at 12 per cent. per annum?

Ans. $3527·453.

3. What is due on a pension of $150 a year, payable half-yearly, but forborne 2 years, allowing half-yearly compound interest, at 6 per cent. per annum?

Ans. $313·772.

4. What is due on a pension of $350 a year, payable quarterly, but forborne 2½ years, allowing quarterly compound interest, at 12 per cent.? *Ans.* $1003·088.

Questions under this rule may be easily wrought by the following table, which shows the amount of an annuity of $1, or £1, forborne for any number of years not exceeding 30, at 3, 4, 5, and 6 per cent., compound interest.

TABLE OF ANNUITIES.

Years.	3 per cent.	4 per cent.	5 per cent.	6 per cent.
1	1·000000	1·000000	1·000000	1·000000
2	2·030000	2·040000	2·050000	2·060000
3	3·090900	3·121600	3·152500	3·183600
4	4·183627	4·246464	4·310125	4·374616
5	5·309136	5·416323	5·525631	5·637093
6	6·468410	6·632975	6·801913	6·975319
7	7·662462	7·898294	8·142008	8·393838
8	8·892336	9·514226	9·249109	9·897468
9	10·159106	10·582795	11·026564	11·491316
10	11·463879	12·006107	12.577893	13·180795
11	12·807796	13·486351	14·206787	14·971643
12	14·192030	15·025805	15·917127	16·869941
13	15·617790	16·626838	17·712983	18·882138
14	17·086324	18·291911	19·598632	21·015066
15	18·598914	20·023588	21·578564	23·275970
16	20·156881	21·824531	23·657492	25·672528
17	21·761588	23·697512	25·840366	28·212880
18	23·414435	25·645431	28·132385	30·905653
19	25·116868	27·671229	30·539004	33·759992
20	26·870374	29·778079	33·065954	36·785591
21	28·676486	31·969202	35·719252	39·992727
22	30·536780	34·247970	38·505214	43·392290
23	32·452884	36·617889	41·430475	46·995828
24	34·426470	39·082604	44·501999	50·815577
25	36·459264	41·645908	47·727099	54·864512
26	38·553042	44·311745	51·113454	59·156383
27	40·709634	47·084214	54·669126	63·705766
28	42·930923	49·967583	58·402583	68·528112
29	45·218850	52·966286	62·322712	73·639798
30	47·575416	56·084938	66·438847	79·058186

5. What is due on a pension of $1000, which has been forborne 27 years, at 3 per cent., compound interest?

From the preceding table, we find the amount of an annuity of $1 for 27 years, at 3 per cent., to be $40·709634, which, multiplied by 1000, gives $40709·634 for the amount due.

6. What is the amount of an annuity of $50, which has been forborne 30 years, at 6 per cent., compound interest? *Ans.* $3952·909.

7. What is the amount of a pension of $300, which has been forborne 19 years, at 5 per cent., compound interest? *Ans.* $9161·701.

8. What is the amount of a pension of $900, which has been forborne 17 years, at 4 per cent., compound interest? *Ans.* $21327·761.

9. What is the amount of an annuity of $75, which has been forborne 13 years, at 5 per cent., compound interest? *Ans.* $1328·474.

Case II.

To find the present worth of an annuity which is to terminate in a given number of years.

The present worth of an annuity is obviously such a sum of money as will, at compound interest, produce an amount equal to the *amount of the annuity.* Therefore, if we find the amount of the annuity by Case I., we may consider it as the amount of a certain principal, which principal is the same as the present worth. We have already been taught how to find the present worth, by rule under *Compound Discount.* Hence, we have this

RULE.

First find the amount of the annuity, as if it were in arrears for the whole time, by the aid of the table under Case I. of ANNUITIES.

Then find the present worth of this amount for the given time and rate per cent., by the use of the table under COMPOUND DISCOUNT.

EXAMPLES.

1. What is the present worth of an annuity of $500, to continue 10 years, interest being 6 per cent.?

By the table under Case I., of Annuities, we find the amount of an annuity of $1 for 10 years, at 6 per cent., to be $13·180795; this, multiplied by 500, gives $6590·3975 for the amount of the annuity.

Now, by the table under Compound Discount, we find the present worth of $1 for 10 years, at 6 per cent., to be $0·558395; which, multiplied by 6590·3975, gives $3680·045 for the present worth required.

2. What is the present worth of an annuity of $100, to continue 20 years, at 5 per cent. interest?

Ans. $1246·222.

The work under this rule may be very much simplified by the use of the following table, which gives the present worth of an annuity of $1, or £1, for any number of years not exceeding 30, at 3, 4, 5, and 6 per cent.

TABLE OF ANNUITIES.

Years.	3 per cent.	4 per cent.	5 per cent.	6 per cent.
1	0·970874	0·961538	0·952381	0·943396
2	1·913470	1·086095	1·859410	1·833393
3	2·828611	2·775091	2·723248	2·673012
4	3·717098	3·629895	3·545950	3·465106
5	4·579707	4·451822	4·329477	4·212364
6	5·417191	5·242137	5·075692	4·917324
7	6·230283	6·002055	5·786373	5·582381
8	7·019692	6·732745	6·463213	6·209794
9	7·786109	7·435332	7·107822	6·801692
10	8·530203	8·110896	7·721735	7·360087
11	9·252624	8·760477	8·306414	7·886875
12	9·954004	9·385074	8·863252	8·383844
13	10·634955	9·985648	9·393573	8·852683
14	11·296070	10·563122	9·898641	9·294984
15	11·936935	11·118387	10·379658	9·712249
16	12·561102	11·652296	10·837770	10·105895
17	13·166118	12·165669	11·274066	10·477260
18	13·753513	12·659197	11·689587	10·827603
19	14·323799	13·133839	12·085321	11·158116
20	14·877475	13·590326	12·462216	11·469921
21	15·415024	14·029160	12·821153	11·764077
22	15·936917	14·451115	13·163003	12·041582
23	16·443608	14·856842	13·488574	12·303379
24	16·935542	15·246963	13·798642	12·550358
25	17·413418	15·622080	14·093945	12·783356
26	17·876842	15·982769	14·375185	13·003166
27	18·327031	16·329586	14·643034	13·210534
28	18·764108	16·663063	14·898127	13·406164
29	19·188455	16·983715	15·141074	13·590721
30	19·600441	17·292033	15·372451	13·764831

To find the present worth of an annuity by means of this table, we must take from it the present worth of $1 for the given time and rate per cent., and multiply it by the number of dollars in the given annuity.

3. What is the present worth of an annuity of $27 for 9 years, at 4 per cent.?

From the table, we find the present worth of $1 for 9 years, at 4 per cent., to be $7·435332; this, multiplied by 27, gives $200·754 for the present worth.

4. What is the present worth of a pension of $75 for 15 years, at 5 per cent.? *Ans.* $778·474.

5. A young man purchases a farm for $924, and agrees to pay for it in the course of 7 years, paying $\frac{1}{7}$ part of the price at the end of each year. Allowing interest to be 6 per cent., how much cash in advance will pay the debt? *Ans.* $736·874.

6. Allowing interest to be 6 per cent., how much shall I gain by paying $15 a year for 10 years, in order to cancel a debt of $160, now due? *Ans.* $49·599.

7. What is the present worth of an annuity of $375 for 13 years, interest being reckoned at 4 per cent.?

Since 375 is $\frac{3}{8}$ of 1000, we may multiply the tabular number by 3, divide by 8, and remove the decimal point three places to the right.

Operation.

```
       9·985648
              3
     8)29·956944
Ans. $3744·618
```

8. What is the present worth of an annuity of $875 for 11 years, interest being 6 per cent.?

Since 875 is $\frac{7}{8}$ of 1000, we may multiply by 7, divide by 8, and remove the decimal point three places to the right.

Operation.

$$\begin{array}{r} 7{\cdot}886875 \\ 7 \\ \hline 8)55{\cdot}208125 \\ \hline \textit{Ans.}\ \$6901{\cdot}015625 \end{array}$$

Or, we might have subtracted $\frac{1}{8}$ of the tabular number from itself, and then have removed the decimal point three places to the right, as in this second

Operation.

$$\begin{array}{r} 8)7{\cdot}886875 \\ 985859375 \\ \hline \textit{Ans.}\ \$6901{\cdot}015625 \end{array}$$

The student ought to exercise himself in seeking short and expeditious methods whenever the nature of the operation will admit of contractions.

NOTE.—When an annuity does not commence until a given time has elapsed, or some particular event has happened, it is called a REVERSION.

Case III.

To find the present worth of an annuity in reversion.

RULE.

Find, by the use of the table under last Case, the present worth of the annuity from the present time up to the end of its continuance; find, also, by the same table, its value for the time before it commences; the difference of these results will be the present worth.

Or, which is the same thing:

Take the difference of the tabular numbers for these two periods, and multiply by the number of dollars in the annuity.

EXAMPLES.

1. What is the present worth of an annuity of $200, to be continued 5 years, but not to commence till 2 years hence, interest being 6 per cent.?

By our table, we find the present worth of $1 for 7 years, at 6 per cent., to be $5·582381; the same for 2 years, is $1·833393; the difference is $3·748988, which, multiplied by 200, gives $749·798 for the present worth.

2. A father leaves to his son a rent of $310 per annum for 8 years, and the reversion of the same rent to his daughter for 14 years thereafter. What is the present worth of the legacy of each, at 6 per cent.?

Operation.

```
    6·209794=tabular number for 8 years.
         310
    --------
    62097940
   18629382
  -----------
  $1925·036140=present worth of son's.
        12·041582=tabular number 8+14=22 years.
         6·209794=tabular number for 8 years.
         --------
         5·831788
              310
          --------
          58317880
         17495364
        -----------
        $1807·854280=present worth of daughter's.
```

3. What is the present worth of a reversion of $100

a year, to commence in 4 years, and to continue for 10 years, interest being at 6 per cent.? *Ans.* \$582·988.

4. What is the present worth of a reversion of \$800 a year, to continue 7 years, but not to commence until the end of 8 years, interest being 4 per cent.?

Ans. \$3508·514.

When the annuity is to continue for ever, it is obvious that its present worth will be that sum whose interest for 1 year is equal to the annuity; therefore, to find the present worth of an annuity to continue for ever, we must divide the annuity by the interest of \$1 for 1 year, at the given rate per cent.

5. How much must be paid at present for the title to an annuity of \$1000, to commence in 7 years, and to continue for ever, interest at 6 per cent.?

Dividing \$1000 by \$0·06, we get, for the present worth, if entered upon immediately, \$16666·66⅔. From table under Compound Discount, we find the present worth of \$1 for 7 years, at 6 per cent., to be \$0·665057; this, multiplied by 16666·66⅔, gives \$11084·283 for the present worth of \$16666·66⅔, which is evidently the same as the present worth of the annuity.

6. What is the present worth of a reversion of \$100 a year, to commence in 4 years, and to continue for ever, interest being 6 per cent.?

Dividing \$100 by \$0·06, we get 1666⅔ for the number of dollars in the present worth, if entered upon immediately.

From the table under Compound Discount, we find the present worth of \$1 for 4 years, at 6 per cent., to be \$0·792094, which must be multiplied by 1666⅔. But, since 1666⅔ is $\frac{5}{3}$ of 1000, we may multiply by 5, divide by 3, and remove the decimal point three places to the right, as in the following

Operation.

$0·792074
5
3)3960470
Ans. $1320·15⅔

---

71. The following tables, which have been computed by the aid of *logarithms*, are added more for curiosity than for any view to their utility.

This table gives the time required for a given principal to double itself, at compound interest, the interest being compounded *yearly*.

Per cent.	Years.	Per cent.	Years.	Per cent.	Years.
1	69·666	4	17·673	7	10·245
1½	46·556	4½	15·748	7½	9·585
2	35·004	5	14·207	8	9·006
2½	28·071	5½	12·946	8½	8·497
3	23·450	6	11·896	9	8·043
3½	20·150	6½	11·007	9½	7·638

The following table gives the time required for a given principal to double itself at compound interest, the interest being compounded *half-yearly*.

Per cent.	Years.	Per cent.	Years.	Per cent.	Years.
1	69·487	4	17·502	7	10·075
$1\frac{1}{2}$	46·382	$4\frac{1}{2}$	15·576	$7\frac{1}{2}$	9·914
2	34·830	5	14·036	8	8·837
$2\frac{1}{2}$	27·899	$5\frac{1}{2}$	12·775	$8\frac{1}{2}$	8·346
3	23·278	6	11·725	9	7·874
$3\frac{1}{2}$	19·977	$6\frac{1}{2}$	10·836	$9\frac{1}{2}$	7·468

This table gives the time required for a given principal to double itself at compound interest, the interest being compounded *quarter-yearly*.

Per cent.	Years.	Per cent.	Years.	Per cent.	Years.
1	69·400	4	17·415	7	9·989
$1\frac{1}{2}$	46·298	$4\frac{1}{2}$	15·490	$7\frac{1}{2}$	9·328
2	34·743	5	13·946	8	8·751
$2\frac{1}{2}$	27·812	$5\frac{1}{2}$	12·686	$8\frac{1}{2}$	8·241
3	23·191	6	11·639	9	7·788
$3\frac{1}{2}$	19·890	$6\frac{1}{2}$	10·750	$9\frac{1}{2}$	7·383

The following table gives the time required for a given principal to double itself at compound interest, the interest being compounded *every instant*.

Per cent.	Years.	Per cent.	Years.	Per cent.	Years.
1	69·315	4	17·329	7	9·902
$1\frac{1}{2}$	46·210	$4\frac{1}{2}$	15·403	$7\frac{1}{2}$	9·242
2	34·657	5	13·863	8	8·665
$2\frac{1}{2}$	27·726	$5\frac{1}{2}$	12·603	$8\frac{1}{2}$	8·155
3	23·105	6	11·552	9	7·702
$3\frac{1}{2}$	19·804	$6\frac{1}{2}$	10·664	$9\frac{1}{2}$	7·296

The following table gives the amount of $1, or £1, for any number of years up to 30, for 5 and 6 per cent., compound interest, the interest being compounded *every instant.*

Years.	5 per cent.	6 per cent.	Years.	5 per cent.	6 per cent.
1	1·0513	1·0618	16	2·2255	2·6116
2	1·1052	1·1275	17	2·3396	2·7731
3	1·1618	1·1972	18	2·4595	2·9446
4	1·2214	1·2712	19	2·5857	3·1267
5	1·2840	1·3498	20	2·7182	3·3201
6	1·3498	1·4333	21	2·8576	3·5253
7	1·4190	1·5219	22	3·0041	3·7433
8	1·4918	1·5161	23	3·1581	3·9748
9	1·5683	1·7160	24	3·3201	4·2206
10	1·6487	1·8221	25	3·4903	4·4815
11	1·7332	1·9348	26	3·6693	4·7587
12	1·8221	2·0544	27	3·8573	5·0529
13	1·9155	2·1771	28	4·0550	5·3653
14	2·0137	2·3163	29	4·2630	5·6971
15	2·1169	2·4596	30	4·4815	6·0492

If we compute the *instantaneous* compound interest at 6·76587 per cent., it will, at the end of the year, be equal to the simple interest at 7 per cent.

In the same way, the *instantaneous* compound interest, at 5·8269 per cent., is the same as simple interest at 6 per cent.

[For some curious results in regard to Instantaneous Compound Interest, see an article which I prepared for the *American Journal of Science*, Vol. 47, No. 1.]

CHAPTER VIII.

BANKING.

72. A Bank is an incorporated institution, created for the purpose of loaning money, receiving deposits, and dealing in exchange.

The *Stock*, or amount of money in trade, is limited by law, and owned by various individuals, who are called *stockholders*.

Banks are allowed to make notes, which are denominated *bank bills*, which circulate as money, because they are obliged to redeem them with *specie*.

It is customary for banks, in most cases, when they loan money, to take the interest in advance;* that is, to deduct it from the face of the note at the time the money is lent. The note is then said to be *discounted*.

The sum to be discounted, or the face of the note, is called the *amount*.

The interest deducted is called the *discount*.

What remains is called the *present worth*, or *proceeds*.

A note to be discounted, or bankable, must be made payable at some future time, and to the order of some person who indorses it.

It is usual for the banks to take interest for three days more than the time specified in the note; and the borrower is not obliged to make payment till those three

* This method of discounting bank notes is usurious, and is fast going out of use, and instead thereof the banks now deduct the discount as found by rule under Art. 66.

days have expired, which are, for this reason, called *days of grace.*

To find the banking discount on any sum of money, we have this

RULE.

Compute the interest (by Case III., Art. **63**,) *on the given sum for three days more than is specified.*

EXAMPLES.

1. What is the banking discount on $1000 for 3 months, at 7 per cent.?

In this example, we find the interest on $1 for 3 months and 3 days, at 6 per cent., to be $0·0155, which, multiplied by 1000, gives $15·50 for the discount at 6 per cent.; this, increased by its sixth part, becomes $18·08$\frac{1}{3}$ for the discount at 7 per cent., as required.

2. What is the banking discount of $150 for 6 months, at 6 per cent.? *Ans.* $4·575.

3. What is the banking discount of $375 for 3 months and 9 days, at 7 per cent.? *Ans.* $7·438.

4. What is the banking discount of $400 for 9 months, at 7 per cent.? *Ans.* $21·233.

5. What is the banking discount of $29·30 for 7 months, at 5 per cent.? *Ans.* $0·867.

6. What is the banking discount of $472 for 10 months, at 7 per cent.? *Ans.* $27·808.

When the present worth of a bankable note, the time for which it is to be discontinued, and the rate per cent. is given, to find the amount, we have this

RULE.

Compute the banking discount on $1 *for the given time and rate per cent.; subtract this discount from* $1, *then divide the present worth by the remainder, and the quotient will be the number of dollars in the amount.*

EXAMPLES.

1. What must be the amount of a bankable note, so that, when discounted for 3 months, at 6 per cent., it shall give a present worth of $600?

In this example, we find the banking discount on $1 for 3 months, to be $0·0155, which, subtracted from $1, gives $0·9845; ∴ dividing $600 by $0·9845, we obtain 609·446 for the number of dollars in the required amount of the note.

2. What must be the face of a bankable note, so that, when discounted for 2 months, at 7 per cent., the borrower shall receive $50? *Ans.* $50·62.

The following table gives the amount of a bankable note, so that, when discounted at 5, 6, or 7 per cent., for any number of months from 1 to 12, the present worth shall be just $1.

Months.	5 per cent.	6 per cent.	7 per cent.
1	1·004604	1·005530	1·006458
2	1·008827	1·010611	1·012402
3	1·013085	1·015744	1·018416
4	1·017380	1·020929	1·024503
5	1·021711	1·026167	1·030662
6	1·026079	1·031460	1·036896
7	1·030485	1·036807	1·043206
8	1·034929	1·042095	1·049593
9	1·039411	1·047669	1·056059
10	1·043932	1·053186	1·062605
11	1·048493	1·058761	1·069233
12	1·053093	1·064396	1·075944

We will now work some examples by the aid of the preceding table :

3. What must be the face of a bankable note, so that, when discounted for 10 months, at 5 per cent., the present worth may be $1000 ?

Looking in the table, directly under the 5 per cent., and adjacent to 10 months, we find $1·043932 ; this, multiplied by 1000, gives $1043·932 for the face of the note required.

4. What must be the face of a bankable note, so that, when discounted for 7 months, at 7 per cent., the present worth may be $70·50 ? *Ans.* $73·546.

5. What amount must I make my note, so that, when discounted at the bank for 12 months, at 7 per cent., I may receive $100 ? *Ans.* $107·594.

6. What must be the amount of a note, so that, when discounted at the bank for 6 months, at 6 per cent., the borrower may receive $365 ? *Ans.* $376·483.

7. What must be the amount of a note, so that, when discounted at the bank for 9 months, at 7 per cent., the borrower may receive $500 ? *Ans.* $527·03.

---

73. The banks, by this method of discounting, obtain a larger per cent. for their money than is obtained by the usual method of loaning money. To illustrate this, suppose A gets a note of $1 discounted at the bank for 12 months, or 1 year, at 7 per cent., he receives $0·93 ; the $0·07 is retained by the bank, it being the interest of \$1 for 1 year. This \$0·07 may now be loaned to B, and its interest again withheld; and so on, for an indefinite period of terms. Hence, at the end of the year, the bank will receive for its \$1, the number of dollars expressed by the sum of the terms of the following geometrical progression:

$1+\frac{7}{100}+(\frac{7}{100})^2+(\frac{7}{100})^3+$, &c.; this, summed, disregarding the 3 days of grace, gives $\frac{100}{93}=1·0752688$. Therefore, in this case, the bank receives 7·52688 *per cent. per annum* for its money.

The longer the time for which they discount, the larger per cent. do they receive.

To make this appear obvious, suppose a person wished his note discounted at the bank for $14\frac{2}{7}$ years, at 7 per cent. In this case, the interest would equal the whole face of the note; so that the bank would withhold the whole amount, be that ever so large, and the borrower would not receive a single cent, but would, nevertheless, be obliged to pay to the bank, at the end of $14\frac{2}{7}$ years, the face of the note. In this case, the per cent. would be *infinite*.

If we go one step farther, and endeavor to discount a note at the bank for a longer period than $14\frac{2}{7}$ years at 7 per cent., we shall be obliged to pay to the bank money from our own pocket before they would accept our note.

The following table shows the per cent. received by banks, when their notes are renewed at the end of any number of months from 1 to 12, at 5, 6, and 7 per cent., *lawful interest*.

Months.	5 per cent.	6 per cent.	7 per cent.
1	5·138	6·200	7·272
2	5·149	6·216	7·295
3	5·160	6·232	7·317
4	5·172	6·248	7·339
5	5·183	6·264	7·362
6	5·194	6·281	7·385
7	5·205	6·298	7·408
8	5·217	6·315	7·432
9	5·228	6·332	7·456
10	5·239	6·349	7·480
11	5·251	6·366	7·503
12	5·263	6·383	7·527

NOTE.—Were it possible for banks to renew their notes *every instant*, the respective rates per cent. would be 5·127, 6·182, and 7·251. This is the same as would be received if the interest were *added every instant*.

CHAPTER IX.

INVOLUTION.

74. INVOLUTION is the method of finding the powers of numbers.

We have already defined the power of a number to be the result arising from multiplying it into itself continually, until the number has been used as a factor as many times as there are units in the exponent denoting the power. Thus, to obtain the cube, or third power of 7, we must use it as a factor three times, which will produce $7 \times 7 \times 7 = 343$.

EXAMPLES.

1. What is the square of 23? *Ans.* 529.
2. What is the cube of 17? *Ans.* 4913.
3. What is the 5th power of 47? *Ans.* 229345007.
4. What is the 9th power of 9? *Ans.* 387420489.
5. What is the square of 22667121? *Ans.* 513798374428641.
6. What is the square of 0·75? *Ans.* 0·5625.
7. What is the cube of 0·65? *Ans.* 0·274625.
8. What is the square of $8\frac{1}{2}$? *Ans.* $72\frac{1}{4}$.

EVOLUTION.

75. Evolution is the reverse of *Involution.* It explains the method of resolving a number into equal factors, which factors are called *roots.*

When a number is resolved into two equal factors, this factor is called the *square root* of the number.

When a number is resolved into three equal factors, the factor is called the *cube root* of the number.

The operation of resolving a number into two equal factors is called the *extraction of the square root.*

EXTRACTION OF THE SQUARE ROOT.

76. If we square 48 by the usual rule, we get $48^2=2304$. But if, instead of 48, we use $40+8$, we shall find, by actual multiplication,

$$\begin{array}{r} 40+8 \\ 40+8 \\ \hline 320+64 \\ 1600+320 \\ \hline 48^2=1600+640+64 \end{array}$$

Now, to reverse this operation, that is, to extract the square root of $1600+640+64$, we proceed as follows:

We take the square root of 1600, which is 40; this is the first part of the root; its square being subtracted from $1600+640+64$, leaves the remainder $640+64$. We see that 640, divided by twice 40, or 80, gives 8 for a quotient, which is the second part of the root required.

Case I.

From the preceding process, we deduce the following rule for the extraction of the square root of a whole number:

RULE.

I. *Point off the given number into periods of two figures each, counting from the right towards the left. When the number of figures is odd, it is evident that the left-hand, or first period, will consist of but one figure.*

II. *Find the greatest square in the first period, and place its root at the right of the number, in the form of a quotient figure in division. Subtract the square of this root from the first period, and to the remainder annex the second period; the result will be the* FIRST DIVIDEND.

III. *Double the root already found, and place it on the left of the number for the* FIRST TRIAL DIVISOR. *See how many times this trial divisor, with a cipher annexed, is contained in the dividend; the quotient figure will be the second figure of the root: this must be placed at the right of the* TRIAL DIVISOR; *the result will be the* TRUE DIVISOR. *Multiply the true divisor by this second figure of the root, and subtract the product from the dividend, and to the remainder annex the next period for a* SECOND DIVIDEND.

IV. *To the last* TRUE DIVISOR *add the last figure of the root for a new* TRIAL DIVISOR, *and continue to operate as before, until all the periods have been brought down.*

EXAMPLES.

1. What is the square root of 531441?

Operation.

		53′14′41(729, root.
	7	49
First trial divisor,	14	414, first dividend.
First true divisor,	142	284
Second trial divisor,	144	13041, second dividend.
Second true divisor,	1449	13041
		0

2. What is the square root of 11390625?

Operation.

3	11′39′06′25)3375
63	9
667	239
6745	189
	5006
	4669
	33725
	33725
	0

In the first example, we exhibited the trial divisors, as well as the true divisors; but in the second example, we adhered more closely to our rule, and placed the succeeding figures of the root at the right of the trial divisors, without again writing them down.

3. What is the square root of 11019960576?

Operation.

```
1         1'10'19'96'05'76(104976
204       1
          ----
3089       1019
20987       816
           -----
209946     20396
           18801
           ------
            159505
            146909
            -------
             1259676
             1259676
             -------
                   0
```

4. What is the square root of 16983563041?

Ans. 130321.

5. What is the square root of 79792266297612001?

Ans. 282475249.

6. What is the square root of 852891037441?

Ans. 923521.

7. What is the square root of 61917364224?

Ans. 248832.

8. What is the square root of 13422659310152401?

Ans. 115856201.

9. The sum of the four numbers, 386, 2114, 3970, 10430, is a perfect square; so also is the sum of any two of these numbers. What are these seven roots?

Ans. 130, 50, 66, 104, 78, 112, 120.

Case II.

To extract the square root of a decimal fraction, or of a number consisting partly of a whole number, and partly of a decimal value, we have this

RULE.

I. *Annex one cipher, if necessary, so that the number of decimals shall be even.*

II. *Then point off the decimals into periods of two figures each, counting from the unit's place towards the right. If there are whole numbers, they must be pointed off as in Case I. Then extract the root as in Case I.*

NOTE.—If the given number has not an exact root, there will be a remainder after all the periods have been brought down, in which case the operation may be extended by forming new periods of ciphers.

EXAMPLES.

1. What is the square root of 3486·784401?
Ans. 59·049.

2. What is the square root of 25·62890625?
Ans. 5·0625.

3. What is the square root of 6·5536? *Ans.* 2·56.

4. What is the square root of 0·00390625?
Ans 0·0625.

5. What is the square root of 17?
Ans. 4·123, nearly.

6. What is the square root of 37·5?
Ans. 6·123, nearly.

7. What is the square root of 0·0000012321?
Ans. 0·00111.

8. What is the square root of 0·0011943936?
Ans. 0·03456.

9. What is the square root of 60·481729?
Ans. 7·777.

Case III.

To extract the square root of a vulgar fraction, or mixed number, we have this

RULE.

I. *Reduce the vulgar fraction, or mixed number, to its simplest fractional form.*

II. *Then extract the square root of the numerator and denominator separately, when they have exact roots; but when they have not, reduce the fraction to a decimal, and proceed as in Case II.*

EXAMPLES.

1. What is the square root of $\frac{256}{324}$? *Ans.* $\frac{8}{9}$.
2. What is the square root of $\frac{10125}{12005}$? *Ans.* $\frac{45}{49}$.
3. What is the square root of $4\frac{21}{25}$? *Ans.* $2\frac{1}{5}$.
4. What is the square root of $\frac{5}{9}$ of $\frac{4}{5}$ of $\frac{4}{7}$ of $\frac{7}{9}$? *Ans.* $\frac{4}{9}$.
5. What is the square root of $4\frac{1}{9}$? *Ans.* 2·027, nearly.
6. What is the square root of $\frac{11}{17}$? *Ans.* 0·8044, nearly.
7. What is the square root of $\frac{13}{49}$? *Ans.* 0·515, nearly.
8. What is the square root of $\frac{1}{365}$? *Ans.* 0·052, nearly.
9. What is the square root of $\frac{1234}{4321}$? *Ans.* 0·534, nearly.
10. What is the square root of $\frac{123344}{447780}$? *Ans.* 0·524, nearly.

Case IV.

When there are many figures required in the root, we may, after obtaining one more than half the number required, find the rest by dividing the remainder by the last *true divisor*, deprived of its right-hand figure. This division should be performed according to the abridged method, as explained under Art. **41.**

EXAMPLES.

1. What is the square root of 11 to 16 decimals?

Operation.

3	11(3·3166247903553998.
63	9
661	2
6626	189
66326	11
663322	661
6633244	439
66332487	39756
66332494|9	4144
	397956
	16444
	1326644
	317756
	26532976
	5242624
	464327409
	59934991
	5969924541
	23574559
	19899748
	3674811
	3316625
	358186
	331662
	26524
	19900
	6624
	5970
	654
	597
	57
	53
	4

In the preceding example, after obtaining 9 figures of the root, by the usual rule, we had, for the remainder, 23574559; the last true divisor was 66332494|9, when deprived of its right-hand figure. We then divided this remainder by this divisor, according to the method of *abridged divisions of decimals*, Art. **41**, and obtained the remaining 8 figures of the root.

2. What is the square root of 3 to 10 decimals?

Ans. 1·7320508076.

3. What is the square root of 0·00008876684 to 10 places of decimals? *Ans.* 0·0094216155.

4. What is the square root of 0·88670811113724 to 10 places of decimals? *Ans.* 0·9416517994.

5. What is the square root of 3·14159265 to 8 places of decimals? *Ans.* 1·77245385.

6. What is the square root of 2 to 9 places of decimals? *Ans.* 1·414213562.

7. What is the square root of 10 to 15 places of decimals? *Ans.* 3·162277660168379.

8. What is the square root of the decimal 0·4444, &c., or of the simple repetend $0{\cdot}\dot{4}$?

Ans. 0·666, &c., or $\frac{2}{3}$.

From this example, we see that the square root of a repetend may also be a repetend.

EXAMPLES

INVOLVING THE

PRINCIPLES OF THE SQUARE ROOT.

77. A TRIANGLE is a figure having three sides, and, consequently, three angles.

When one of the angles is right, like the corner of a square, the triangle is called a *right-angled triangle.* In this case, the side opposite the right angle is called the *hypothenuse.*

It is an established proposition of geometry, that the square of the hypothenuse is equal to the sum of the squares of the other two sides.

From the above proposition, it follows that the square of the hypothenuse, diminished by the square of one of the sides, equals the square of the other side.

By means of these properties, when two sides of a right-angled triangle are given, the third side can be found.

EXAMPLES.

1. How long must a ladder be to reach the top of a house 40 feet high, when the foot of it is 30 feet from the house?

In this example, it is obvious that the ladder forms the hypothenuse of a right-angled triangle, whose sides are 30 and 40 feet, respectively. Therefore, the square of the length of the ladder must equal the sum of the squares of 30 and 40.

$$30^2 = 900$$
$$40^2 = 1600$$

$\sqrt{2500} = 50 =$ the length of the ladder.

2. Suppose a ladder 100 feet long be placed 60 feet from the foot of a tree; how far up the tree will the top of the ladder reach? *Ans.* 80 feet.

3. Two persons start from the same place, and go, the one due north 50 miles, the other due west 80 miles. How far apart are they? *Ans.* 94·34 miles, nearly.

4. What is the distance through the opposite corners of a square yard? *Ans.* 4·24264 feet, nearly.

5. The distance between the lower ends of two equal rafters, in the different sides of a roof, is 32 feet, and the height of the ridge, above the foot of the rafters, is 12 feet. What is the length of a rafter? *Ans.* 20 feet, nearly.

6. What is the distance measured through the centre of a cube, from one corner to its opposite corner, the cube being 3 feet, or 1 yard, on a side? *Ans.* 5·196 feet.

We know, from the principles of geometry, that all similar surfaces, or areas, are to each other as the squares of their like dimensions.

7. Suppose we have two circular pieces of land, the one 100 feet in diameter, the other 20 feet in diameter; how much more land is there in the larger than in the smaller?

By the above principle of geometry, it follows that the quantity of land in the two circles must be as the squares of the diameters; that is, as 100^2 to 20^2, or as 25 to 1. Hence, there is 25 times as much in the one piece as there is in the other.

8. Suppose two persons, the one 6 feet high, the other 5 feet, to be both well proportioned in all respects; how much more cloth will it take to make a suit of clothes for the first, than for the second?

Ans. { It will require $1\frac{11}{25}$ times as much for the first as for the second.

9. Suppose, by observation, it is found that 4 gallons of water flow through a circular orifice of 1 inch in diameter, in one minute; how many gallons would, under similar circumstances, be discharged through an orifice of 3 inches in diameter, in the same length of time? *Ans.* 36 gallons.

10. What must be the circumference of a circular pond, which shall contain $\frac{1}{16}$ part as much surface as a pond $13\frac{1}{2}$ miles in circumference? *Ans.* $3\frac{3}{8}$ miles.

11. Required the width and depth of a rectangular box, whose length is 3 feet, which shall contain 30000 solid inches; the width being to the depth as 2 to 3.

A box, whose length is 3 feet $=36$ inches, width 2 inches, and depth 3 inches, must contain $36\times2\times3$ inches.

Then, $\frac{30000}{1}\times\frac{1}{36}\times\frac{1}{2}\times\frac{1}{3}=\frac{1250}{9}$, whose square root is $\frac{25}{3}\sqrt{2}=11{\cdot}785$, nearly; this, multiplied respectively by 2 and 3, will give

Ans. $\begin{cases}11{\cdot}785\times2=23{\cdot}57 \text{ inches.}\\ 11{\cdot}785\times3=35{\cdot}36 \quad \text{"}\end{cases}$

Or, this question may be solved by the following method:

Had the width of the box been the same as the depth, its volume would have been one half more than it now is; that is, would have been $\frac{3}{2}$ of $30000=45000$ cubic inches, which, divided by 36 inches, the length, will give 1250 square inches for the end area of this new box; or, which is the same thing, 1250 is the square of the depth of the original box. Hence, $\sqrt{1250}=35{\cdot}36$, nearly, for the number of inches in the depth. The width is $\frac{2}{3}$ of $35{\cdot}36=23{\cdot}57$, nearly.

12. What length of thread is required to wind spirally around a cylinder 2 feet in circumference, and 3 feet in length, so as to go but once around?

It is evident that if the cylinder be developed, or placed upon a plane, and caused to roll once over, that the convex surface of the cylinder will give a rectangle, whose width is 2 feet, and length 3 feet; at the same time the thread will form its diagonal. Hence, the length of the thread is $\sqrt{4+9}=\sqrt{13}=3{\cdot}60555$ feet.

13. Seven men purchase a grindstone, of 60 inches in diameter. What part of the diameter must each grind off, so as to have $\frac{1}{7}$ of the whole stone?

Solution.

In this question, we disregard the thickness of the stone.

After the first one has ground off his share, the remaining stone will be $\frac{6}{7}$ of the original stone. Therefore, its diameter will be $60\sqrt{\frac{6}{7}}=\frac{60}{7}\sqrt{42}=55{\cdot}54921$, nearly.

The diameter, after the second one has ground off his share, will be $60\sqrt{\frac{5}{7}}=\frac{60}{7}\sqrt{35}=50{\cdot}70925$, nearly.

The diameter, after the third one has ground off his share, will be $60\sqrt{\frac{4}{7}}=\frac{60}{7}\sqrt{28}=45{\cdot}35574$, nearly.

The diameter, after the fourth one has ground off his share, will be $60\sqrt{\frac{3}{7}}=\frac{60}{7}\sqrt{21}=39{\cdot}27922$, nearly.

The diameter, after the fifth one has ground off his share, will be $60\sqrt{\frac{2}{7}}=\frac{60}{7}\sqrt{14}=32{\cdot}07135$, nearly.

The diameter, after the sixth one has ground off his share, will be $60\sqrt{\frac{1}{7}}=\frac{60}{7}\sqrt{7}=22{\cdot}67787$, nearly.

Hence, the parts of the diameter ground off are as follows:

				Inches, nearly.
The 1st	ground	off	60·00000 — 55·54921 =	4·45079
2d	"	"	55·54921 — 50·70925 =	4·83996
3d	"	"	50·70925 — 45·35574 =	5·35351
4th	"	"	45·35574 — 39·27922 =	6·07652
5th	"	"	39·27922 — 32·07135 =	7·20787
6th	"	"	32·07135 — 22·67787 =	9·39348
7th	"	"	22·67787	= 22·67787

EXTRACTION OF THE CUBE ROOT.

78. If we cube 45 by the usual process, we find $45^3 = 91125$.

If, instead of 45, we take its equal, 40+5, and then cube it by actual multiplication, as explained under Art. **4**, we shall have this

Operation.

$$
\begin{array}{r}
45 = 40 + 5 \\
40 + 5 \\
\hline
200 + 25 \\
1600 + 200 \\
\hline
45^2 = 1600 + 400 + 25 \\
40 + 5 \\
\hline
8000 + 2000 + 125 \\
64000 + 16000 + 1000 \\
\hline
45^3 = 64000 + 24000 + 3000 + 125
\end{array}
$$

Now, to reverse this process, that is, to extract the cube root of 64000+24000+3000+125, we proceed as follows:

I. We find the cube root of 64000 to be 40, which we place to the right of the number, in the form of a quotient in division, for the first part of the root sought.

We also place it on the left of the number in a column headed 1st Col.; we next multiply it into itself, and place the result in a column headed 2d Col.; this last result, being multiplied by 40, gives 64000, which we subtract from the number, and obtain the remainder 24000+3000+125, which we will call the *first dividend.*

II. We obtain the second term of the 1st column by adding the first term to itself; the result being multiplied by this first term, and added to the first term of the 2d column, gives its second term. Again, adding this first term to the second term of the 1st column, we get its third term.

III. We seek how many times the second term of the 2d column is contained in the first dividend; or, simply how many times it is contained in its first part, 24000, which gives 5 for the second part of the root.

IV. Finally, we add this 5 to the last term of the 1st column, whose result, multiplied by 5, and added to the last term of the 2d column, gives its third term; which, multiplied by 5, gives 27125 = 24000 + 3000 + 125.

1st Col.	2d Col.	*Number.*	*Root.*
40	1600	64000 + 24000 + 3000 + 125	(40 + 5.
80	4800	64000	
120	5425	24000 + 3000 + 125 = 27125	
125		5425 × 5 = 27125	
		0	

This work can be written in a more condensed form, as follows, where the ciphers upon the right have been omitted.

1st Col.	2d Col.	*Number.*	*Root.*
		91125	(45
4	16	64	
8	48	27125	
12	5425	27125	
125		0	

Case I.

From the preceding operation, we may draw the following rule for extracting the cube root of a whole number.

RULE.

I. *Since the cube of any number cannot have more than three times as many places of figures as the number, we must separate the number into periods of three figures each, counting from the unit's place towards the left. When the number of figures is not divisible by* 3, *the left-hand period will contain less than three figures.*

II. *Seek the greatest cube of the first, or left-hand period; place its root at the right of the number, after the manner of a quotient in division; also place it to the left of the number for the first term of a column marked* 1st COL. *Then multiply it into itself, and place the product for the first term of a column marked* 2d COL. *Again, multiply this last result by the same figure, and subtract the product from the first period, and to the remainder annex the next period, and it will give the* FIRST DIVIDEND. *This same figure must be added to the first term of the* 1st *column; the sum will be its second term, which must be multiplied by the same figure, and the product added to the first term of the* 2d *column; this sum will be its second term, which we shall name the* FIRST TRIAL DIVISOR.

The same figure of the root must be added to the second term of the 1st *column, to form its third term.*

III. *See how many times the trial divisor, with two*

ciphers annexed, is contained in the dividend; the quotient figure will be the second figure of the root, which must be placed at the right of the first figure; also annex it to the third term of the 1st column, and multiply the result by this second figure, and add the product, after advancing it two places to the right, to the last term of the 2d column. Again, multiply this last result by this second figure of the root, and subtract the product from the dividend, and to the remainder annex the next period for a NEW DIVIDEND.

Proceed with this second figure of the root precisely as was done with the first figure, and so continue, until all the periods have been brought down.

NOTE.—This rule may be readily deduced from the rule for extracting the cube root of a polynomial, as given in my ALGEBRA.

EXAMPLES.

1. Extract the cube root of 387420489.

Operation.

1st Col.	2d Col.	Number.	Root.
		387′420′489(	729
7	49	343	
14	147, 1st trial divisor.	44420, 1st div.	
212	15124	30248	
214	15552, 2d trial divisor.	14172489, 2d div.	
2169	1574721	14172489	
		0	

Explanation.

The greatest cube of the first period, 387, is 343, whose root is 7, which we place to the right of the number for the first figure of the root sought. We also place it for the first term of the first column; which, multiplied into itself, gives $7\times7=49$, for the first term of the 2d column, which, in turn, multiplied by 7, gives $49\times7=343$, which, subtracted from the first period, 387, leaves the remainder 44, to which, annexing the next period, 420, we get 44420 for the *first dividend.*

Again, adding 7 to the first term, 7, of the 1st column, we get $7+7=14$, for the second term of the 1st column, which, multiplied by 7, gives $14\times7=98$; this, added to the first term of the second column, gives 147 for the second term of the 2d column, or the *first trial divisor.*

Again, adding 7 to the second term of the 1st column, we get $14+7=21$, for the third term of the first column.

The *trial divisor*, with two ciphers annexed, becomes 14700, which is contained 3 times in the *first dividend*, 44420. Since the *trial divisor* is less than the true divisor, it will sometimes give too large a quotient figure; such is the case in this present example, where 2 is the second figure of the root.

This second figure, 2, of the root, annexed to the third term of the 1st column, gives 212; which, multiplied by 2, gives 424, which, being advanced two places to the right, must be added to 147, the last term of the 2d column. The sum 15124 will form the third term of the 2d column, which, multiplied by 2, gives $15124\times2=30248$, which, subtracted from the first dividend, leaves 14172 for the remainder, to which, annexing the next period, 489, we get 14172489 for the *second dividend.*

Again, to the last term, 212, of the 1st column, adding 2, we get 214 for the next term; which, multiplied by

2, gives 428, which, added to 15124, gives 15552 for the *second trial divisor*. Again, adding 2 to 214, we get 216 for the fifth term of the 1st column.

The *second trial divisor*, with two ciphers annexed, becomes 1555200, which is contained 9 times in the *second dividend*, 14172489; therefore, 9 is the third figure of the root, which, annexed to 216, gives 2169 for the last term of the first column, which, multiplied by 9, gives 19521, which, advanced two places to the right, and then added to 15552, gives 1574721; this, multiplied by 9, gives 14172489, which, subtracted from the *second dividend*, leaves no remainder.

2. What is the cube root of 913517247483640899?

Operation.

1st Col.	2d Col.	*Number.*	*Root.*
		913′517′247′483′640′899(	970299
9	81	729	
18	243	184517	
277	26239	183673	
284	28227	844247483	
29102	282328204	564656408	
29104	282386412	279591075640	
291069	28241260821	254171347389	
291078	28243880523	25419728251899	
2910879	2824414250211	25419728251899	
		0	

3. What is the cube root of 10077696? *Ans.* 216.

4. What is the cube root of 2357947691? *Ans.* 1331.

5. What is the cube root of 42875? *Ans.* 35.

6. What is the cube root of 117649? *Ans.* 49.

7. What is the cube root of 350356403707485209?
Ans. 704969.

8. What is the cube root of 75084686279296875?
Ans. 421875.

9. What is the cube root of 7256313856?
Ans. 1936.

10. What is the cube root of 106868920913284608?
Ans. 474552.

11. The four following numbers,

2080913082956455142636,	4937801347510680732948,
7262810476410016163052,	2149721086932415893409 48,

when added together, by taking two at a time, produce six distinct sums, each of which is a *perfect cube.* What are the six roots of these cube numbers?

Ans. { 19146344, 21062342, 60097344,
23021160, 60359866, 60571840.

Case II.

To extract the cube root of a decimal fraction, or of a number consisting partly of a whole number, and partly of a decimal value, we have this

RULE.

I. *Annex ciphers to the decimals, if necessary, so that the whole number of decimal places may be divisible by* 3.

II. *Separate the decimals into periods of three figures each, counting from the decimal point towards the right, and proceed as in whole numbers.*

Note.—If the given number has not an exact root, there will be a remainder after all the periods have been brought down. The process may be continued by annexing ciphers for new periods.

EXAMPLES.

1. What is the cube root of 0·469640998917? *Ans.* 0·7773.
2. What is the cube root of 18·609625? *Ans.* 2·65.
3. What is the cube root of 1·25992105? *Ans.* 1·08005974, nearly.
4. What is the cube root of 2? *Ans.* 1·25992105, nearly.
5. What is the cube root of 3? *Ans.* 1·442249, nearly.
6. What is the cube root of 1860867? *Ans.* 123.

Case III.

To extract the cube root of a vulgar fraction, or mixed number, we have this

RULE.

I. *Reduce the fraction, or mixed number, to its simplest fractional form.*

II. *Extract the cube root of the numerator and denominator separately, if they have exact roots; but when they have not, reduce the fraction to a decimal, and then extract the root by Case II.*

EXAMPLES.

1. What is the cube root of $\frac{2197}{4913}$? *Ans.* $\frac{13}{17}$.
2. What is the cube root of $\frac{85169}{1770723}$? *Ans.* $\frac{23}{29}$.
3. What is the cube root of $17\frac{1}{8}$? *Ans.* 2·577, nearly.

4. What is the cube root of $5\frac{1}{7}$?

Ans. 1·726, nearly.

5. What is the cube root of $\frac{81}{99}$?

Ans. 0·9353, nearly.

6. What is the cube root of $\frac{2}{3}$?

Ans. 0·8736, nearly.

7. What is the cube root of $47\frac{1}{3}$?

Ans. 3·6173, nearly.

8. What is the cube root of $101\frac{1}{7}$?

Ans. 4·6592, nearly.

9. What is the cube root of $9\frac{3}{8}$?

Ans. 2·1085, nearly.

Case IV.

When there are many decimal places required in the root, we may, after obtaining one more decimal figure than half the required number, find the rest by dividing the remainder by the last term of the *second column*.

Before dividing, we can omit from the right of the divisor so many figures as to leave but one more than the number of additional figures required in the root, observing to omit from the right of the dividend one figure less than was omitted in the divisor. The division must then be performed according to the abridged method, as explained under Art. **41.**

EXAMPLES.

1. What is the cube root of 7, carried to 9 decimal places?

Operation.

1st Col.	2d Col.	*Root.* 7(1·912931182
1	1	1
2	3	6
39	651	5859
48	1083	141
571	108871	108871
572	109443	32129
5732	10955764	21911528
5734	10967232	10217472
57369	1097239521	9875155689
57378	1097755923	342316311
573873	10977|7313919	329331941757
		12984|369243
		10978
		2006
		1098
		908
		878
		30
		22
		8

In this example, we proceed in the usual way, until we obtain 1·91293; the remainder is 12984369243; the last term of the second column is 109777313919; therefore, we obtain four more figures by dividing 12984369243 by 109777313919; but these four figures may be obtained with equal accuracy by dividing 12984 by 10977, which gives the remaining figures 1182.

2. Extract the cube root of $\frac{1}{4}=0·25$ to 13 decimal places.

Operation.

1st Col.	2d Col.	*Root.*
		0·250(0·6299605249474
6	36	216
12	108	34
182	11164	22328
184	11532	11672
1869	1170021	10530189
1878	1186923	1141811
18879	118862211	1069759899
18888	119032203	72051101
188976	11904354156	71426124936
188982	11905488048	624976064
18898805	1190549 | 74974025	595274874870125
		2970118 | 9129875
		2381099
		589019
		476220
		112799
		107149
		5650
		4762
		888
		833
		55
		47
		8

In this example, after obtaining seven decimal figures in the root, by the usual process, the remainder was 29701189129875, and the last term in the *second column* was 119054974974025; and, since we wish but six figures by division, we reject seven figures from the right of the remainder, and eight figures from the right of the term of the *second column*, and then divide by the rule for abridging the work, Art. **41**, and obtain the remaining figures of the root.

3. Extract the cube root of 9 to 9 decimals

Operation.

```
                                             Root.
1st Col.     2d Col.                  9(2·080083823
  2            4                      8
                                      ‾
  4           12                      1
608           124864                   998912
                                       ──────
616           129792                     1088
624008        12979|6992064              1038375936512
                                         ─────────────
                                          49624|063488
                                          38939
                                          ─────
                                          10685
                                          10383
                                          ─────
                                            302
                                            259
                                            ───
                                             43
                                             39
                                             ──
                                              4
```

4. What is the cube root of $15\frac{2}{3}$ to 5 decimal places?

Ans. 2·50222.

5. What is the cube root of $\frac{1}{3·14159265}$ to 8 decimals?

Ans. 0·68278406.

6. What is the cube root of 0·0000031502374 to 13 decimals?

Ans. 0·0146593403377.

7. What is the cube root of $\frac{1}{2}$ to 21 decimals?

Ans. 0·793700525984099737376.

8. What is the cube root of $\frac{1}{7}$ to 10 decimals?

Ans. 0·5227579585.

9. What is the cube root of $\frac{14}{19}$ to 7 decimals?

Ans. 0·9032157.

10. What is the cube root of $\frac{1}{365}$ to 8 decimals?

Ans. 0·13992727.

EXAMPLES

INVOLVING THE

PRINCIPLES OF THE CUBE ROOT.

79. It is an established theorem of geometry, that all similar solids are to each other as the cubes of their like dimensions.

1. If a cannon ball 3 inches in diameter weigh 8 pounds, what will a ball of the same metal weigh, whose diameter is 4 inches?

By the above theorem, we have $3^3 : 4^3 :: 8$ pounds : $18\frac{26}{27}$ pound, for the answer.

2. Suppose the diameter of the sun to be 887681 miles; the diameter of the earth, 7912 miles. How many times greater in bulk is the sun than the earth?

$(887681)^3 = 699472706450842241$;

$(7912)^3 = 495289174528$;

$699472706450842241 \div 495289174528 = 1412251$ times, nearly.

3. How many cubic quarter inches can be made out of a cubic inch? *Ans.* 64.

4. Required the dimensions of a rectangular box, which shall contain 20000 solid inches; the length, breadth, and depth being to each other as 4, 3, and 2.

$\frac{20000}{1}\times\frac{1}{4}\times\frac{1}{3}\times\frac{1}{2}=\frac{2500}{3}$, whose cube root is $5\sqrt[3]{\frac{20}{3}}=$ 9·4103, nearly.

$$Ans.\begin{cases} 9·4103\times4=37·6412, \text{ length.} \\ 9·4103\times3=28·2309, \text{ breadth.} \\ 9·4103\times2=18·8206, \text{ depth.} \end{cases}$$

Or, as follows:

If we were to augment the width of this box, so as to make it as wide as it is long, its volume would become $\frac{4}{3}$ of $20000=26666\frac{2}{3}$. Again, if we augment the depth of this new box, so that it may be as deep as it is wide, and as it is long, its volume will become 2 times $26666\frac{2}{3}$ $=53333\frac{1}{3}$, which is the contents of a cubical box, whose side is equal to the length of the original box. Hence, $\sqrt[3]{53333\frac{1}{3}}=37·641$, nearly, for the length. The width is $\frac{3}{4}$ of this length, and the depth is $\frac{1}{2}$ this length.

5. What is the side of a cube which will contain as much as a chest 8 feet 3 inches long, 3 feet wide, and 2 feet 7 inches deep? *Ans.* 47·9843 inches.

6. Four ladies purchased a ball of exceeding fine thread, 3 inches in diameter. What portion of the diameter must each wind off so as to share of the thread equally?

Solution.

After the first one had wound off her share, the ball which remained would contain $\frac{3}{4}$ as much thread as it did in the first place. Therefore, its diameter was $3\sqrt[3]{\frac{3}{4}}=\frac{3}{2}\sqrt[3]{6}=2·72568$ inches, nearly.

The diameter, after the second one had wound off her share, was $3\sqrt[3]{\frac{2}{4}}=\frac{3}{2}\sqrt[3]{4}=2·38110$ inches, nearly.

The diameter, after the third one had wound off her share, was $3\sqrt[3]{\frac{1}{4}}=\frac{3}{2}\sqrt[3]{2}=1{\cdot}88988$ inches, nearly.

Hence, the portions of the diameter which they must wind off are as follows:

						Inches, nearly.
The 1st lady	must	wind	off	$3{\cdot}00000-2{\cdot}72568$	$=$	$0{\cdot}27432$
2d	"	"	"	$2{\cdot}72568-2{\cdot}38110$	$=$	$0{\cdot}34458$
3d	"	"	"	$2{\cdot}38110-1{\cdot}88988$	$=$	$0{\cdot}49122$
4th	"	"	"	$1{\cdot}88988$	$=$	$1{\cdot}88988$

ROOTS OF ALL POWERS.

80. WHENEVER the index denoting the root required is a composite number, the root can be found by successive extractions of the roots denoted by the prime factors of the original index.

Thus, the 4th root may be found by extracting the 2d root twice in succession.

The 6th root may be obtained by extracting the 3d root of the second root.

The 8th root may be found by extracting the 2d root three times in succession.

When the index denoting the root is a prime, we must have some direct method of obtaining the root.

By a similar train of reasoning, as was used in deducing the *rule for the cube root*, we determine, in general, for any root, the following

RULE.

I. *Point the number off into periods of as many figures each as there are units in the index denoting the root.*

II. *Find, by trial, the figure of the first period, which will be the first figure of the root; place this figure to the left, in a column called the* FIRST COLUMN. *Then multiply it by itself, and place the product for the first term of the* SECOND COLUMN. *This, multiplied by the same figure, will give the first term of the* THIRD COLUMN. *Thus continue until the number of columns is one less than the units in the index denoting the root.*

Multiply the term in the LAST COLUMN *by the same figure, and subtract the product from the first period, and to the remainder bring down the next period, and it will form the* FIRST DIVIDEND.

Again, add this same figure to the term of the FIRST COLUMN, *multiply the sum by the same figure, and add the product to the term of the* SECOND COLUMN; *which, in turn, must be multiplied by the same figure, and added to the term of the* THIRD COLUMN, *and so on, till we reach the* LAST COLUMN, *the term of which will form the* FIRST TRIAL DIVISOR.

Again, beginning with the FIRST COLUMN, *repeat the above process until we reach the column next to the last. And so continue to do, until we obtain as many terms in the* FIRST COLUMN *as there are units in the index denoting the root; observing, in each successive operation, to terminate on the column of the next inferior order.*

III. *Seek how many times the* FIRST TRIAL DIVISOR, *when there are annexed to it as many ciphers, less one, as there are units in the index, is contained in the* FIRST DIVIDEND; *the quotient figure will be the second figure of the root. Then proceed with this figure the same as was done with the first figure; observing to advance the terms of the different columns as many places to the right as the number expressing the order of the column; that is, advancing the terms of the* FIRST COLUMN *one place, those of the* SECOND COLUMN *two places, and so for the succeeding columns.*

After completing the requisite number of terms in the different columns, by means of this second figure of the root, then proceed to obtain the third figure of the root in the same way as the second figure was obtained; and in this way the operation can be continued until all the periods are brought down. If there is still a remainder, the process can be extended by forming periods of ciphers.

EXAMPLES.

1. What is the fifth root of 36936242722357?

Operation.

1st Col.	2d Col.	3d Col.	4th Col.	*Root.*
5	25	125	625	3693′62427′22357(517
10	75	500	3125	3125
15	150	1250	32525251	56862427
20	250	1275251	33826005	32525251
251	25251	1300754	347673946051	2433717622357
252	25503	1326510		2433717622357
253	25756	1344842293		0
254	26010			
2557	2618899			

2. What is the 7th root of 12311715481324O9344 ?

Operation.

1st Col.	2d Col.	3d Col.	4th Col.	5th Col.	6th Col.	*Root.*
						12311′7154813′2409344(384
3	9	27	81	243	729	2187
6	27	108	405	1458	5103	101247154813
9	54	270	1215	5103	11568197824	92545582592
12	90	540	2835	808149728	21076554688	87015722212409344
15	135	945	37231216	1188544608	21753930553102336	87015722212409344
18	189	1110152	47549360	1663938528		0
218	20644	1289768	59424240	169343966275584		
226	22452	1484360	72979760			
234	24324	1694440	737528368896			
242	26260	1920520				
250	28260	1932692224				
258	30324					
2664	3043056					

3. What is the eleventh root of 11 ?

[The work in this question is so lengthy, we have been compelled to make use of two pages in our operation.]

Operation.

1st Col.	2d Col.	3d Col.	4th Col.	5th Col.	6th Col.	7th Col.
1	1	1	1	1	1	1
2	3	4	5	6	7	8
3	6	10	15	21	28	36
4	10	20	35	56	84	120
5	15	35	70	126	210	330
6	21	56	126	252	462	4438023168
7	28	84	210	462	569011584	5822306304
8	36	120	330	53505792	692141568	7488297728
9	45	165	3652896	61564992	832995712	9474862720
10	55	176448	4029600	70427072	993282496	11824496640
112	5724	188352	4431040	80143392	1174816960	123951454696751104
114	5952	200720	4858160	90767232	1379524608	129850860737626112
116	6184	213560	5311920	102353824	1426622074187776	
118	6420	226880	5793296	114960384	1474851510218752	
120	6660	240688	6303280	11774366546944		
122	6904	254992	6842880	12057359007744		
124	7152	269800	69582036736			
126	7404	285120	70748115200			
128	7660	288309184				
130	7920	291519616				
1324	797296					
1328	802608					

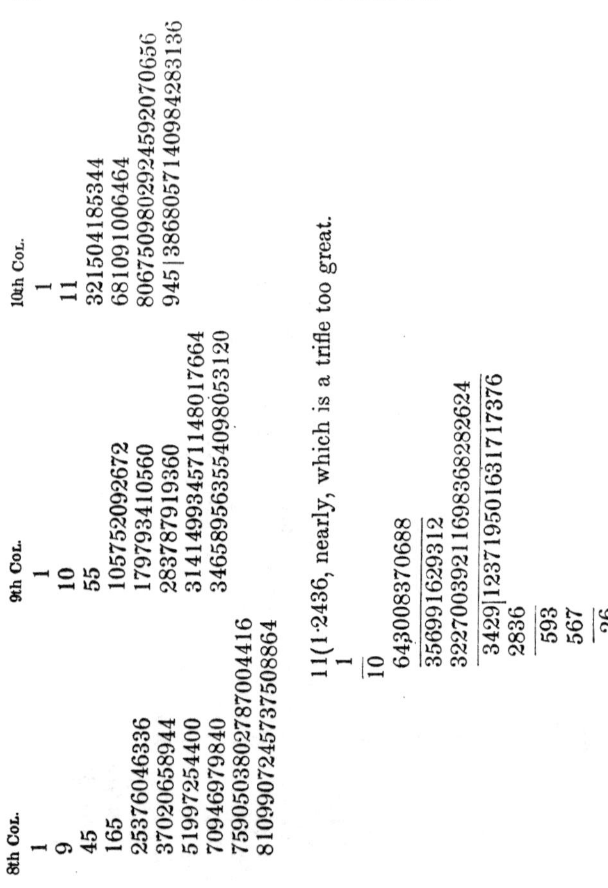

8th Col.	9th Col.	10th Col.
1	1	1
9	10	11
45	55	321504185344
165	105752092672	681091006464
25376046336	179793410560	8067509802924592070656
37020658944	283787919360	945|3868057140984283136
51997254400	3141499345711480017664	
70946979840	346589563554098053120	
7590503802787004416		
8109907245737508864		

```
11(1·2436, nearly, which is a trifle too great.
 1
10
 64,3008370688
 356991629312
 32270039211698368282624
  3429|123719501631717376
  2836
   593
   567
    26
```

4. What is the fifth root of 5?

Ans. 1·37974, nearly.

5. What is the seventh root of 2?

Ans. 1·10409, nearly.

CHAPTER X.

ARITHMETICAL PROGRESSION.

81. A SERIES of numbers which succeed each other regularly by a common difference, is said to be in *arithmetical progression.*

When the terms are constantly increasing, the series is an *arithmetical progression ascending.*

When the terms are constantly decreasing, the series is an *arithmetical progression descending.*

Thus, 1, 3, 5, 7, 9, &c., is an ascending arithmetical progression; and 10, 8, 6, 4, 2, is a descending arithmetical progression.

In arithmetical progression, there are five things to be considered:

1. *The first term.*
2. *The last term.*
3. *The common difference.*
4. *The number of terms.*
5. *The sum of all the terms.*

These quantities are so related to each other, that any three of them being given, the remaining two can be found.

If we denote the five things by the numerals 1, 2, 3, 4, 5, they may be taken by threes, as follows:

1, 2, 3,	giving	4	and	5,	making	2 cases.
1, 2, 4	"	3	"	5	"	2 "
1, 2, 5	"	3	"	4	"	2 "
1, 3, 4	"	2	"	5	"	2 "
1, 3, 5	"	2	"	4	"	2 "
1, 4, 5	"	2	"	3	"	2 "
2, 3, 4	"	1	"	5	"	2 "
2, 3, 5	"	1	"	4	"	2 "
2, 4, 5	"	1	"	3	"	2 "
3, 4, 5	"	1	"	2	"	2 "

Hence, there must be 20 distinct cases arising from the different combinations of these five quantities.

To give a demonstration to all the rules of these 20 cases, would be a very difficult task for the ordinary processes of arithmetic; we will, therefore, content ourselves with demonstrating a few of the most important of them.

Case I.

By our definition of an ascending arithmetical progression, it follows that the second term is equal to the first, increased by the common difference; the third is equal to the first, increased by twice the common difference; the fourth is equal to the first, increased by three times the common difference; and so on, for the succeeding terms.

Hence, when we have given the first term, the common difference, and the number of terms, to find the last term, we have this

RULE.

To the first term add the product of the common difference into the number of terms, less one.

EXAMPLES.

1. What is the 100th term of an arithmetical progres sion, whose first term is 2, and common difference 3?

In this example, the number of terms, less one, is 99; which, multiplied by the common difference, 3, gives 297, which, added to the first term, 2, makes 299 for the 100th term.

2. What is the 50th term of the arithmetical progression, whose first term is 1, the common difference being $\frac{1}{2}$? *Ans.* $25\frac{1}{2}$.

3. A man buys 10 sheep, giving $1 for the first, $3 for the second, $5 for the third, and so increasing in arithmetical progression. What will the last sheep cost at that rate? *Ans.* $19.

4. A person bought 100 yards of cloth; he gave 2 s. 6 d. for the first yard, 2 s. 10 d. for the second yard, 3 s. 2 d. for the third yard, and so continued to give 4 d. more for each yard than he gave for the preceding one. How much did he give for the last yard?

Ans. £1 15 s. 6 d.

Case II.

From the nature of an arithmetical progression, we see that the second term added to the next to the last term, is equal to the first added to the last, since the second term is as much greater than the first, as the next to the last is less than the last. After the same method of reasoning, we infer that the sum of any two terms equidistant from the extremes is equal to the sum of the extremes.

Hence, it follows that the terms will average just half the sum of the extremes.

Therefore, when we have given the first term, the last term, and the number of terms, to find the sum of all the terms, we have this

RULE.

Multiply half the sum of the extremes by the number of terms.

EXAMPLES.

1. The first term of an arithmetical progression is 2, the last term is 50, and the number of the terms is 17. What is the sum of all the terms?

In this example, half the sum of the extremes is $\frac{2+50}{2}=26$; this, multiplied by the number of terms, gives $26\times 17=442$, for the sum required.

2. The first term of an arithmetical progression is 13, the last term is 1003, and the number of terms is 100. What is the sum of the progression? *Ans.* 50800.

3. A person travels 25 days, going 11 miles the first day, and 135 the last day; the miles which he traveled in the successive days form an arithmetical progression. How far did he go in the 25 days? *Ans.* 1825 miles.

4. Bought 7 books, the prices of which are in arithmetical progression. The price of the first was 8 shillings, and the price of the last was 28 shillings. What did they all come to? *Ans.* £6 6 s.

Case III.

By Case I. we see that the last term is equal to the first term, increased by the product of the common difference into the number of terms, less one.

Hence, the first term must equal the last term, diminished by the product of the common difference into the number of terms, less one.

Therefore, when we have given the last term, the number of terms, and the common difference, to find the last term, we have this

RULE.

From the last term subtract the product of the common difference into the number of terms, less one.

EXAMPLES.

1. The last term of an arithmetical progression is 375, the common difference is 7, and the number of terms is 54. What is the first term?

In this example, the common difference, multiplied by the number of terms, less one, is $7 \times 53 = 371$, which, subtracted from the last term, gives $375 - 371 = 4$, for the first term.

2. The last term of an arithmetical progression is $39\frac{1}{3}$, the common difference is $\frac{2}{3}$, and the number of terms is 59. What is the first term? *Ans.* $\frac{2}{3}$.

3. A note is paid in 15 annual instalments; the payments are in arithmetical progression, whose common difference is 3; the last payment is 49 dollars. What is the first payment? *Ans.* $7.

Case IV.

From Case I. we see that the product of the common difference into the number of terms, less one, is equal to the last term diminished by the first. Therefore, the difference of the last and first terms, divided by the common difference, is equal to the number of terms, less one.

Hence, when we have given the first term, the last term, and the common difference, to find the number of terms, we have this

RULE.

Divide the difference of the extremes by the common difference, and to the quotient add one.

EXAMPLES.

1. The first term of an arithmetical progression is 5, the last term 176, and the common difference is 3. What is the number of terms?

In this example, the difference of the extremes is $176-5=171$; this, divided by the common difference, gives $\frac{171}{3}=57$; which, increased by 1, becomes 58, for the number of terms required.

2. The first term of an arithmetical progression is 11, the last term is 88, and the common difference is 7. What is the number of terms? *Ans.* 12.

3. A note becomes due in annual instalments, which are in arithmetical progression, whose common difference is 3; the first payment is 7 dollars, the last payment is 49 dollars. What is the number of instalments? *Ans.* 15.

Case V.

We learn from Case I. that the product of the com mon difference into the number of terms, less one, is equal to the last term diminished by the first. Therefore, the difference of the last and first terms, divided by the number of terms, less one, will give the common difference.

Hence, when we have given the first term, the last term, and the number of terms, to find the common difference, we have this

RULE.

Divide the difference of the extremes by the number of terms, less one.

EXAMPLES.

1. The first term of an arithmetical progression is 5, the last term is 176, and the number of terms 58. What is the common difference?

In this example, the difference of the extremes is 171; which, divided by the number of terms, less one, becomes $\frac{171}{57}=3$, for the common difference.

2. A person performs a journey in 17 days; the distances traveled on the successive days were in arithmetical progression; the first day he went 4 miles, and the last day he went 84. How many miles more did he go on each day, than on the preceding day? *Ans.* 5 miles.

3. A man has 7 sons, whose ages are in arithmetical progression; the age of the eldest is 41 years, the youngest is 5 years old. How many years is the common difference of their ages? *Ans.* 6 years.

Case VI.

By Case II. we know that the sum of all the terms of an arithmetical progression is equal to half the sum of the extremes multiplied into the number of terms; therefore, the number of terms is equal to the sum of all the terms divided by half the sum of the extremes.

Hence, when we have given the first term, the last term, and the sum of all the terms, to find the number of terms, we have this

RULE.

Divide the sum of all the terms by half the sum of the extremes.

EXAMPLES.

1. The first term of an arithmetical progression is 1, the last term is 1001, and the sum of all the terms is 251001. What is the number of terms?

In this example, half the sum of the extremes is $\frac{1001+1}{2}=501$; then, dividing the sum of all the terms by this, we obtain $\frac{251001}{501}=501$, for the number of terms.

2. In a triangular field of corn, the number of hills in the successive rows are in arithmetical progression; in the first row there is but one hill, in the last row there are 81 hills, and the whole number of hills in the field is 1681. How many rows are there? *Ans.* 41.

3. A man bought a certain number of yards of cloth, for $152·50, giving 4 cents for the first yard, and in creasing regularly on each succeeding yard, up to the

last yard, for which he gave $3·01. How many yards of cloth did he purchase? *Ans.* 100.

Case VII.

We also infer from Case II. that the sum of all the terms, divided by half the number of terms, will give the sum of the extremes. Therefore, if from the quotient of the sum of all the terms, divided by half the number of terms, we subtract the last term, we shall have left the first term.

Hence, when we have given the last term, the number of terms, and the sum of all the terms, to find the first term, we have this

RULE.

From the quotient of the sum of all the terms, divided by half the number of terms, subtract the last term.

EXAMPLES.

1. If the last term of an arithmetical progression is 170, the number of terms 50, and the sum of all the terms 4450, what is the first term?

In this example, the sum of all the terms, divided by half the number of terms, is $\frac{4450}{25}=178$; from which, subtracting the last term, we obtain $178-170=8$, for the first term.

2. A person wishes to discharge a debt of $1125 in 18 annual payments, which shall be in arithmetical progression. How much must his first payment be, so as to bring his last payment $120? *Ans.* $5.

3. The miles which a person travels in 19 successive days, form an arithmetical progression, whose last term

is 80, the sum of all the terms 950. How many miles does he travel the first day? *Ans.* 20 miles.

Case VIII.

From what has been said under Case VII. we infer that the first term subtracted from the quotient of the sum of all the terms divided by half the number of terms, will give the last term.

Hence, when we have given the sum of all the terms, the first term, and the number of terms, to find the last term, we have this

RULE.

From the quotient of the sum of all the terms, divided by half the number of terms, subtract the first term.

EXAMPLES.

1. If the first term of an arithmetical progression is 7, the number of terms 1000, and the sum of all the terms 560000, what is the last term?

In this example, the sum of all the terms, divided by half the number of terms, gives $\frac{560000}{500}=1120$; from which subtract the first term, we get $1120-7=1113$, for the last term.

2. If the first term of an arithmetical progression is 7, the number of terms 16, and the sum of all the terms 142, what is the last term? *Ans.* $10\frac{3}{4}$.

3. The first term of an arithmetical progression is 13, the number of terms 100, and the sum of all the terms 50300. What is the last term? *Ans.* 993.

NOTE.—The remaining cases are obtained by combining the conditions of Cases I. and II.

Case XI.

Given the common difference, the number of terms, and the sum of all the terms, to find the first term.

RULE.

Divide the sum of the terms by the number of terms; from this quotient subtract half the product of the common difference into the number of terms, less one.

EXAMPLES.

1. The common difference of the terms of an arithmetical progression is 7, the number of terms 54, and the sum of all the terms is 10233. What is the first term?

In this example, the sum of the terms, divided by the number of terms, is $189\frac{1}{2}$; half the product of the number of terms, less one, into the common difference, is $185\frac{1}{2}$; which, subtracted from $189\frac{1}{2}$, leaves 4 for the first term.

2. The common difference of the terms of an arithmetical progression is $\frac{2}{3}$, the number of terms is 59, and the sum of all the terms is 1180. What is the first term? *Ans.* $\frac{2}{3}$.

3. A father divides $2000 among five sons, so that each should receive $40 more than his next younger brother. What is the share of the youngest?

Ans. $320.

Case X.

Given the common difference, the number of terms, and the sum of all the terms, to find the last term.

RULE.

To the quotient of the sum of the terms, divided by the number of terms, add half the product of the common difference into the number of terms, less one.

EXAMPLES.

1. The common difference of the terms of an arithmetical progression is 6, the number of terms is 7, and the sum of all the terms is 161. What is the last term?

In this example, the sum of the terms, divided by the number of terms, is $\frac{161}{7}=23$. Again, the common difference multiplied into the number of terms, less one, is $6\times 6=36$, the half of which is 18, which, added to 23, gives 41 for the last term.

2. The common difference of the terms of an arithmetical progression is 7, the number of terms is 54, and the sum of all the terms is 10233. What is the last term? *Ans.* 375.

3. The common difference of the terms of an arithmetical progression is 6, the number of terms is 14, and the sum of all the terms is 4970. What is the last term? *Ans.* 394.

Case XI.

Given the first term, the common difference, and the number of terms, to find the sum of all the terms.

RULE.

To twice the first term, add the product of the common difference into the number of terms, less one; multiply this sum by half the number of terms.

EXAMPLES.

1. The first term of an arithmetical progression is 37, the common difference is 11, and the number of terms 99. What is the sum of all the terms?

In this example, the product of the common difference into the number of terms, less one, is $11 \times 98 = 1078$; this, added to twice the first term, gives $74 + 1078 = 1152$, which, multiplied by half the number of terms, gives 57024 for the sum of all the terms.

2. The first term of an arithmetical progression is 7, the common difference is $1\frac{1}{2}$, and the number of terms 37. What is the sum of all the terms? *Ans.* 1258.

3. A person buys 37 sheep, paying for them in arithmetical progression; for the first he gives 3 shillings, and increases 1 shilling for each succeeding one. How much did they all come to? *Ans.* £38 17 s.

Case XII.

Given the first term, the common difference, and the last term, to find the sum of all the terms.

RULE.

Divide the difference of the squares of the last and first terms by twice the common difference, and to this quotient add half the sum of the last and first terms.

EXAMPLES.

1. The first term of an arithmetical progression is 16, the common difference is 2, and the last term 100. What is the sum of all the terms?

In this example, the difference of the squares of the last and first terms is 9744, which, divided by twice the common difference, gives 2436; this, increased by half the sum of the last and first term, becomes 2494, for the sum of all the terms.

2. The first term of an arithmetical progression is 5, the common difference is 7, and the last term is 75. What is the sum of all the terms? *Ans.* 440.

3. The first term of an arithmetical progression is 8, the common difference 3, and the last term 170. What is the sum of all the terms? *Ans.* 4895.

Case XIII.

Given the common difference, the number of terms, and the last term, to find the sum of all the terms.

RULE.

From twice the last term, subtract the product of the common difference into the number of terms, less one, multiply this remainder by half the number of terms.

EXAMPLES.

1. The common difference of the terms of an arithmetical progression is 11, the number of terms is 19, and the last term is 199. What is the sum of all the terms?

In this example, the product of the common difference into the number of terms, less one, is $11 \times 18 = 198$; this, subtracted from twice the last term, gives 200, which, multiplied by half the number of terms, becomes 1900, for the sum of all the terms.

2. The common difference of the terms of an arithmetical progression is 15, the number of terms is 47, and the last term is 545. What is the sum of all the terms? *Ans.* 9400.

3. The common difference of the terms of an arithmetical progression is $4\frac{1}{2}$, the number of terms is 100, and the last term is 1000. What is the sum of all the terms? *Ans.* 77725.

Case XIV.

Given the first term, the number of terms, and the sum of all the terms, to find the common difference.

RULE.

From twice the sum of the terms, subtract twice the product of the first term into the number of terms; divide this remainder by the product of the number of terms into the number of terms, less one.

EXAMPLES.

1. The first term of an arithmetical progression is 21, the number of terms is 50, and the sum of all the terms is 3500. What is the common difference?

In this example, twice the product of the first term into the number of terms, is 2100; which, subtracted from twice the sum of the terms, gives 4900; the number of terms multiplied into the number of terms, less one, is 2450; hence, 4900, divided by 2450, gives 2 for the common difference.

2. The first term of an arithmetical progression is $\frac{5}{7}$, the number of terms is 13, and the sum of all the terms is $139\frac{4}{7}$. What is the common difference? *Ans.* $1\frac{2}{3}$.

3. The first term of an arithmetical progression is $\frac{3}{4}$, the number of terms is 26, and the sum of all the terms is $60\frac{1}{8}$. What is the common difference? *Ans.* $\frac{1}{8}$

Case XV.

Given the first term, the last term, and the sum of all the terms, to find the common difference.

RULE.

Divide the difference of the squares of the last and first term by twice the sum of all the terms, diminished by the sum of the last and first term.

EXAMPLES.

1. The first term of an arithmetical progression is $\frac{5}{7}$, the last term is $20\frac{5}{7}$, and the sum of all the terms is $139\frac{2}{7}$. What is the common difference?

In this example, the difference of the squares of the last and first term is $(20\frac{5}{7})^2-(\frac{5}{7})^2=428\frac{4}{7}$; twice the sum of all the terms, diminished by the sum of the last and first term, is $257\frac{1}{7}$. Dividing $428\frac{4}{7}$ by $257\frac{1}{7}$, we get $\frac{5}{3}=1\frac{2}{3}$, for the common difference.

2. The first term of an arithmetical progression is 8, the last term is 170, and the sum of all the terms is 4895. What is the common difference? *Ans.* 3.

3. The first term of an arithmetical progression is 12, the last term is 102, and the sum of all the terms is 912. What is the common difference? *Ans.* 6.

Case XVI.

Given the number of terms, the last term, and the sum of all the terms, to find the common difference.

RULE.

From twice the product of the number of terms into the last term, subtract twice the sum of all the terms; divide the remainder by the product of the number of terms into the number of terms, less one.

EXAMPLES.

1. The number of terms of an arithmetical progression is 17, the last term is 50, and the sum of all the terms is 442. What is the common difference?

In this example, twice the product of the number of terms into the last term is $2 \times 17 \times 50 = 1700$; the product of the number of terms into the number of terms, less one, is $17 \times 16 = 272$; also 1700, diminished by twice the sum of all the terms, becomes 816, which, divided by 272, gives 3 for the common difference.

2. The number of terms of an arithmetical progression is 14, the last term is 14, and the sum of all the terms is 105. What is the common difference?

Ans. 1.

3. The number of terms of an arithmetical progression is 7, the last term is 41, and the sum of all the terms is 161. What is the common difference?

Ans. 6.

NOTE.—The remaining four cases require for their solution the Extraction of the Square Root.

Case XVII.

Given the first term, the common difference, and the sum of all the terms, to find the number of terms.

RULE.

Subtract the common difference from twice the first term; divide the remainder by twice the common difference; to the square of this quotient add the quotient of twice the sum of all the terms divided by the common difference; extract the square root of the sum; then divide twice the first term, diminished by the common difference, by twice the common difference, and subtract this quotient from the root just found.

EXAMPLES.

1. The first term of an arithmetical progression is 7, the common difference is $\frac{1}{4}$, and the sum of all the terms is 142. What is the number of terms?

In this example, the common difference, subtracted from twice the first term, gives $13\frac{3}{4}$, which, divided by twice the common difference, gives $27\frac{1}{2}$, which, squared, becomes $756\frac{1}{4}$. Twice the sum of all the terms, divided by the common difference, gives 1136, which, added to $756\frac{1}{4}$, gives $1892\frac{1}{4}$, the square root of which is $43\frac{1}{2}$; from this, subtracting $27\frac{1}{2}$, we get 16 for the number of terms.

2. The first term of an arithmetical progression is 2, the common difference is 3, and the sum of all the terms is 442. What is the number of terms? *Ans.* 17.

3. The first term of an arithmetical progression is $\frac{3}{4}$, the common difference is $\frac{1}{8}$, and the sum of all the terms is $60\frac{1}{8}$. What is the number of terms? *Ans.* 26.

Case XVIII.

Given the common difference, the last term, and the sum of all the terms, to find the number of terms.

RULE.

To twice the last term add the common difference; divide the sum by twice the common difference; square the quotient, and from this square subtract the quotient of twice the sum of the terms divided by the common difference; extract the square root of the remainder; then subtract this root from the quotient of the sum of twice the last term and common difference, divided by twice the common difference.

EXAMPLES.

1. The common difference of the terms of an arithmetical progression is $\frac{1}{3}$, the last term is $35\frac{1}{2}$, and the sum of all the terms is 1900. What is the number of terms?

In this example, twice the last term, increased by the common difference, is $71\frac{1}{3}$, which, divided by twice the common difference, gives 107; this, squared, becomes 11449. Again, twice the sum of all the terms, divided by the common difference, gives 11400; this, subtracted from 11449, gives 49, whose square root is 7. Subtracting this root from 107, we get 100 for the number of terms.

2. The common difference of the terms of an arithmetical progression is $\frac{1}{8}$, the last term is $3\frac{7}{8}$, and the sum of all the terms is $60\frac{1}{8}$. What is the number of terms? *Ans.* 26.

3. The common difference of the terms of an arithmetical progression is 1, the last term is 14, and the sum of all the terms is 105. What is the number of terms? *Ans.* 14.

Case XIX.

Given the first term, the common difference, and the sum of all the terms, to find the last term.

RULE.

From the first term subtract half the common difference, and to the square of the remainder add twice the product of the common difference into the sum of all the terms; then extract the square root; which, diminished by half the common difference, will give the last term.

EXAMPLES.

1. The first term of an arithmetical progression is 4, the common difference is 7, and the sum of all the terms is 10233. What is the last term?

In this example, half the common difference, subtracted from the first term, gives $\frac{1}{2}$, which, squared, is $\frac{1}{4}$; this, added to twice the product of the common difference into the sum of all the terms, which is $2\times7\times10233=143262$, gives $\frac{573049}{4}$, whose square root is $\frac{757}{2}$; from this, subtract half the common difference, and we find $\frac{757}{2}-\frac{7}{2}=\frac{750}{2}=375$ for the last term.

2. The first term of an arithmetical progression is $\frac{2}{3}$, the common difference is $\frac{2}{3}$, and the sum of all the terms is 1180. What is the last term? *Ans.* $39\frac{1}{3}$.

3. A man has several sons, whose ages are in arithmetical progression; the age of the youngest is 5 years, the common difference of their ages is 6 years, and the sum of all their ages is 161. What is the age of the eldest? *Ans.* 41 years.

Case XX.

Given the common difference, the last term, and the sum of all the terms, to find the first term.

RULE.

Add half the common difference to the last term, and from the square of the sum subtract twice the product of the common difference into the sum of all the terms; then extract the square root of the remainder, and to this root add half the common difference.

EXAMPLES.

1. The common difference of the terms of an arithmetical progression is 4, the last term is 1008, and the sum of all the terms is 127512. What is the first term?

In this example, half the common difference, added to the last term, gives 1010, which, squared, is 1020100; twice the product of the common difference into the sum of all the terms is 1020096, which, subtracted from 1020100, leaves 4, the square root of which is 2; this, increased by half the common difference, becomes 4 for the first term.

2. The common difference of the terms of an arithmetical progression is 3, the last term is 49, and the sum of all the terms is 420. What is the first term?

Ans. 7.

3. The common difference of the terms of an arithmetical progression is 10, the last term is 1003, and the sum of all the terms is 50800. What is the first term?

Ans. 13.

CHAPTER XI.

GEOMETRICAL PROGRESSION.

82. A SERIES of numbers which succeed each other regularly, by a constant multiplier, is called a *geometrical progression.*

This constant factor, by which the successive terms are multiplied, is called the *ratio.*

When the ratio is greater than a unit, the series is called an *ascending geometrical progression.*

When the ratio is less than a unit, the series is called a *descending geometrical progression.*

Thus, 1, 3, 9, 27, 81, &c., is an ascending geometrical progression, whose ratio is 3.

And 1, $\frac{1}{4}$, $\frac{1}{16}$, $\frac{1}{64}$, &c., is a descending geometrical progression, whose ratio is $\frac{1}{4}$.

In geometrical progression, as in arithmetical progression, there are five things to be considered:

1. *The first term.*
2. *The last term.*
3. *The common ratio.*
4. *The number of terms.*
5. *The sum of all the terms.*

These quantities are so related to each other, that any three being given, the remaining two can be found.

Hence, as in arithmetical progression, it may be shown that there must be 20 distinct cases arising from the different combinations of these five quantities.

The solution of some of these cases requires a knowledge of higher principles of mathematics than can be detailed by arithmetic alone.

We will give a demonstration of the rules of some of the most important cases.

Case I.

By the definition of a geometrical progression, it follows that the second term is equal to the first term multiplied by the ratio; the third term is equal to the first term, multiplied by the second power of the ratio; the fourth term is equal to the first term, multiplied by the third power of the ratio; and so on, for the succeeding terms.

Hence, when we have given the first term, the ratio, and the number of terms, to find the last term, we have this

RULE.

Multiply the first term by the power of the ratio whose exponent is one less than the number of terms.

EXAMPLES.

1. The first term of a geometrical progression is 1, the ratio is 2, and the number of terms is 7. What is the last term?

In this example, the power of the ratio, whose exponent is one less than the number of terms, is $2^6 = 64$, which, multiplied by the first term, 1, still remains 64, for the last term.

2. The first term of a geometrical progression is 5, the ratio is 4, and the number of terms 9. What is the last term? *Ans.* 327680.

3. A person traveling, goes 5 miles the first day, 10 miles the second day, 20 miles the third day, and so on, increasing in geometrical progression. If he continue to travel in this way for 7 days, how far will he go the last day? *Ans.* 320 miles.

Case II.

If we multiply all the terms of a geometrical progression by the ratio, we shall obtain a *new* progression, whose first term equals the second term of the *old* progression; the second term of our *new* progression will equal the third term of the *old* progression, and so on for the succeeding terms. Hence, the sum of the *old* progression, omitting the first term, equals the sum of the *new* progression, omitting its last term. The sum of the *new* progression is equal to the *old* progression repeated as many times as there are units in the ratio. Therefore, the difference between the *new* progression and the *old* progression is equal to the *old* progression repeated as many times as there are units in the ratio, less one. But we also know that the difference between these progressions is equal to the last term of the *new* progression diminished by the first term of the *old* progression; and, since the *new* progression was formed by multiplying the respective terms of the *old* progression by the ratio, it follows that the last term of the *new* progression is equal to the last term of the *old* progression repeated as many times as there are units in the ratio.

Therefore, the last term of the *new* progression, diminished by the first term of the *old* progression, is equal to the last term of the *old* progression repeated as many times as there are units in the ratio and diminished by the first term of the *old* progression. Hence, we finally obtain this condition :

That the sum of all the terms of a geometrical progression, repeated as many times as there are units in the ratio, less one, is equal to the last term multiplied by the ratio and diminished by the first term.

Hence, when we have given the first term of a geometrical progression, the last term, and the ratio, to find the sum of all the terms, we have this

RULE.

Subtract the first term from the product of the last term into the ratio ; divide the remainder by the ratio, less one.

EXAMPLES.

1. The first term of a geometrical progression is 4, the last term is 78732, and the ratio is 3. What is the sum of all the terms?

In this example, the first term, subtracted from the product of the last term into the ratio, is 236192, which, divided by the ratio, less one, gives 118096, for the sum of all the terms.

2. The first term of a geometrical progression is 5, the last term is 327680, and the ratio is 4. What is the sum of all the terms? *Ans.* 436905.

3. A person sowed a peck of wheat, and used the

whole crop for seed the following year; the produce of this second year again for seed the third year, and so on. If, in the last year, his crop is 1048576 pecks, how many pecks did he raise in all, allowing the increase to have been in a four-fold ratio? *Ans.* 1398101 pecks.

Case III.

Since by Case I. the last term is equal to the first term multiplied into a power of the ratio whose exponent is equal to the number of terms, less one, it follows that the first term is equal to the last term divided by the power of the ratio whose exponent is one less than the number of terms.

Hence, when we have given the last term, the ratio, and the number of terms, to find the first term, we have this

RULE.

Divide the last term by a power of the ratio whose exponent is one less than the number of terms.

EXAMPLES.

1. The last term of a geometrical progression is 1048576, the ratio is 4, and the number of terms is 11. What is the first term?

In this example, the ratio, 4, raised to a power whose index is 10, one less than the number of terms, is $4^{10} = 1048576$; $\therefore$ 1048576, divided by 1048576, gives 1 for the first term.

2. A man has 6 sons, among whom he divides his estate in a geometrical progression, whose ratio is 2;

the last son received \$4800. How much did the first son receive? *Ans* \$150.

3. A person bought 10 bushels of wheat, paying for it in geometrical progression, whose ratio is 3; the last bushel cost him \$196·83. What did he give for the first bushel? *Ans.* 1 cent.

Case IV.

We also discover from Case I. that the last term divided by the first term, will give the power of the ratio, whose exponent is the number of terms, less one.

Hence, when we have given the first term, the last term, and the number of terms, to find the ratio, we have this

RULE.

Divide the last term by the first term; extract that root of the quotient which is denoted by the number of terms, less one.

EXAMPLES.

1. The first term of a geometrical progression is 1, the last term is 64, and the number of terms is 7. What is the ratio?

In this example, the last term, divided by the first term, is 64; the number of terms, less one, is 6, ∴ we must extract the 6th root of 64; we first extract the square root, which is 8, we now extract the cube root of 8, which is 2, for the ratio.

2. In a country, during peace, the population increased every year in the same ratio, and so fast that in the space of 5 years it became from 10000 to 14641 souls. By what ratio was the increase, yearly? *Ans.* $\frac{11}{10}$.

3. The first term of a geometrical progression is 4, the last term is 78732, and the number of terms is 10. What is the ratio? *Ans.* 3.

Case V.

If in Case II. we write the product of the first term into the power of the ratio, whose exponent is the number of terms, less one, instead of the last term, as drawn from Case I., we shall have the sum of all the terms, repeated as many times as there are units in the number of terms, less one, equal to the power of the ratio, whose exponent is equal to the number of terms diminished by one, and multiplied by the first term.

Hence, when we have given the first term, the ratio, and the number of terms, to find the sum of all the terms, we have this

RULE.

From the power of the ratio, whose exponent is the number of terms, subtract one, divide the remainder by the ratio, less one, and multiply the quotient by the first term.

EXAMPLES.

1. The first term of a geometrical progression is 3, the ratio is 4, and the number of terms is 9. What is the sum of all the terms?

In this example, the ratio, raised to a power whose exponent is the number of terms, is $4^9 = 262144$; this, diminished by one, becomes 262143, which, divided by 3, gives 87381; this, multiplied by the first term, becomes $87381 \times 3 = 262143$, for the sum of all the terms.

2. A king in India, named SHERAN, wished (according to the Arabic author ASEPHAD,) that SESSA, the inventor of chess, should himself choose a reward. He requested the grains of wheat which arise when 1 is calculated for the first square of the board, 2 for the second square, 4 for the third, and so on; reckoning for each of the 64 squares of the board twice as many grains as for the preceding. When it was calculated, to the astonishment of the king, it was found to be an enormous number. What was it?

Ans. 18446744073709551615 grains.

3. A gentleman married his daughter on New-Year's day, and gave her husband 1 shilling towards her portion, and was to double it on the first day of every month during the year. What was her portion?

Ans. £204 15 s.

Case VI.

We know from Case V. that the sum of all the terms multiplied by the ratio, less one, is equal to one subtracted from the power of the ratio, whose exponent is the number of terms, and this remainder multiplied by the first term.

Hence, when we have given the sum of all the terms, the number of terms, and the ratio, to find the first term, we have this

RULE.

Multiply the sum of all the terms by the ratio, less one; divide the product by the power of the ratio, whose index is the number of terms, after diminishing it by one.

EXAMPLES.

1. The sum of all the terms of a geometrical progression is 262143, the number of terms is 9, and the ratio is 4. What is the first term?

In this example, the sum of all the terms, multiplied by the ratio, less one, is $262143 \times 3 = 786429$; the power of the ratio, whose exponent is the number of terms, is $4^9 = 262144$; this, diminished by 1, becomes 262143; ∴ 786429, divided by 262143, gives 3 for the first term.

2. The sum of all the terms of a geometrical progression is $591\frac{741}{4096}$, the number of terms is 7, and the ratio is $\frac{7}{4}$. What is the first term? *Ans.* 9.

3. If a debt of $4095 is discharged in 12 months by paying sums which are in geometrical progression, the ratio of which is 2, how much was the first payment?

Ans. $1.

Case VII.

We have shown under Case II. that the sum of all the terms, multiplied by the ratio, less one, is equal to the first term subtracted from the last term into the ratio; therefore, the first term is equal to the product of the ratio into the last term, diminished by the product of the ratio, less one, into the sum of all the terms.

Hence, when we have given the sum of all the terms, the last term, and the ratio, to find the first term, we have this

RULE.

Multiply the last term by the ratio, and from the product subtract the product of the sum of all the terms into the ratio, less one.

EXAMPLES.

1. The sum of all the terms of a geometrical progression is 436905, the last term is 327680, and the ratio is 4. What is the first term?

In this example, we find the last term, multiplied by the ratio, to be 1310720. The product of the sum of the terms into the ratio, less one, is 1310715; ∴ 1310720 $-1310715=5$, for the first term.

2. The sum of all the terms of a geometrical progression is 6138, the last term is 3072, and the ratio is 2. What is the first term? *Ans.* 6.

3. The sum of all the terms of a geometrical progression is 1860040, the last term is 1240029, and the ratio is 3. What is the first term? *Ans.* 7.

Case VIII.

From the condition under Case II., we see that the ratio, multiplied into the sum of all the terms, diminished by the last term, is equal to the sum of all the terms, diminished by the first term.

Hence, when we have given the first term, the last term, and the sum of all the terms, to find the ratio, we have this

RULE.

Divide the sum of all the terms, diminished by the first term, by the sum of all the terms, diminished by the last term.

EXAMPLES.

1. The first term of a geometrical progression is 5, the last term is 327680, and the sum of all the terms is 436905. What is the ratio?

In this example, the sum of all the terms, diminished by the first term, is 436900, and the sum of all the terms, diminished by the last term, is 109225; ∴ 436900, divided by 109225, gives 4 for the ratio.

2. The first term of a geometrical progression is 6, the last term is 3072, and the sum of all the terms is 6138. What is the ratio? *Ans.* 2.

3. The first term of a geometrical progression is 7, the last term is 1240029, and the sum of all the terms is 1860040. What is the ratio? *Ans.* 3.

NOTE.—The demonstration of the rules for the four following cases have not been given; they may, however, be obtained by combining the conditions of some of the foregoing cases.

Case IX.

Given the first term, the ratio, and the sum of all the terms, to find the last term.

RULE.

To the first term add the product of the ratio, less one, into the sum of all the terms; divide this sum by the ratio.

EXAMPLES.

1. The first term of a geometrical progression is 4, the ratio is 3, and the sum of all the terms is 118096. What is the last term?

In this example, the product of the ratio, less one, into the sum of all the terms is 236192, which, added to the first term, gives 236196; this, divided by the ratio, gives 78732, for the last term.

2. A man bought a certain number of yards of cloth,

giving 3 cents for the first yard, 6 cents for the second yard, 12 cents for the third yard, and so on, for the succeeding yards. If the whole number of yards cost \$122·63, what did the last cost? *Ans.* \$62·33.

3. A person bought a certain number of pears for £4 5 s. 3 d. 3 far.; he gave 1 farthing for the first, 2 farthings for the second, 4 for the third, and so on, doubling each time. What did he pay for the last?

Ans. £2 2 s. 8 d.

Case X.

Given the ratio, the number of terms, and the sum of all the terms, to find the last term.

RULE.

Raise the ratio to a power whose exponent is the number of terms, less one; multiply together this power, the sum of all the terms, and the ratio, less one; then divide this product by one less than the power of the ratio, whose exponent is the number of terms.

EXAMPLES.

1. The ratio of the terms of a geometrical progression is 3, the number of terms is 10, and the sum of all the terms is 118096. What is the last term?

In this example, the ratio, raised to a power whose exponent is the number of terms, less one, is $3^9=19683$; this, multiplied by the sum of all the terms, and the ratio, less one, is $19683\times118096\times2=4648967136$; the power of the ratio, whose exponent is the number of terms, is 59049; this, diminished by 1, becomes 59048; ∴ 4648967136, divided by 59048, gives 78732, for the last term.

2 The ratio of the terms of a geometrical progression is 3, the number of terms is 10, and the sum of all the terms is 295240. What is the last term?

Ans. 196830.

3. The ratio of the terms of a geometrical progression is 2, the number of terms is 11, and the sum of all the terms is 20470? What is the last term?

Ans. 10240.

Case XI.

Given the first term, the number of terms, and the last term, to find the sum of all the terms.

RULE.

Extract the root denoted by the number of terms, less one, of the last and first terms; then raise these roots to a power, whose exponent is the number of terms; then divide the difference of these powers by the difference of the roots.

EXAMPLES.

1. The first term of a geometrical progression is 1, the number of terms is 10, and the last term is 19683. What is the sum of all the terms?

In this example, we must extract the 9th root of the last and first terms, which give 3 and 1 for the roots; these must each be raised to the 10th power, which give 59049 and 1, the difference of which is 59048; this, divided by $3-1=2$, gives 29524, for the sum of all the terms.

2. The first term of a geometrical progression is 1, the last term is 2048, and the number of terms is 12. What is the sum of all the terms? *Ans.* 4095.

3. The first term of a geometrical progression is 1, the last term is 10077696, and the number of terms is 10. What is the sum of all the terms?

Ans. 12093235.

Case XII.

Given the ratio, the number of terms, and the last term, to find the sum of all the terms.

RULE.

Raise the ratio to a power whose exponent is the number of terms; from this power subtract one, and multiply the remainder by the last term; divide this product by the product of the ratio, less one, into the power of the ratio, whose exponent is the number of terms, less one.

EXAMPLES.

1. The ratio of the terms of a geometrical progression is 2, the number of terms is 12, and the last term is 2048. What is the sum of all the terms?

In this example, the ratio, raised to a power whose exponent is the number of terms, is $2^{12}=4096$; this, diminished by 1, becomes 4095, which, multiplied by 2048, becomes 8386560; again, the power of the ratio, whose exponent is one less than the number of terms, is 2048, which, multiplied by the ratio, less one, is not changed; $\therefore$ 8386560, divided by 2048, gives 4095, for the sum of all the terms.

2. The ratio of the terms of a geometrical progression is $\frac{3}{2}$, the number of terms is 8, and the last term is $106\frac{403}{512}$. What is the sum of all the terms?

Ans. $307\frac{441}{512}$.

3. The ratio of the terms of a geometrical progression is $\frac{7}{4}$, the number of terms is 7, and the last term is $258\frac{2073}{4096}$. What is the sum of all the terms?

Ans. $591\frac{741}{4096}$.

NOTE.—The eight remaining cases in geometrical progression cannot be solved by the ordinary processes of arithmetic, but require for their solution a knowledge of *logarithms*, and *algebraic equations* above the second degree.

---

83. WHEN the ratio of a geometrical progression is less than a unit, the first term will be the largest, and the last term the least; the progression, will, in this case, be descending. But, if we consider the series of terms in a reverse order, that is, calling the last term the first, and the first the last, the progression may then be considered as ascending.

If a decreasing geometrical progression be continued to an infinite number of terms, we may neglect the last term as of no appreciable value; we can find the sum of such a progression by Case II., when it is modified, as follows:

Given the first term of a descending geometrical progression, and the ratio, to find the sum of all the terms, when continued to infinity.

RULE.

Divide the first term by a unit diminished by the ratio.

EXAMPLES.

1. What is the sum of all the terms of the infinite series 1, $\frac{1}{2}$, $\frac{1}{4}$, $\frac{1}{8}$, &c.?

In this example, a unit, diminished by the ratio, is $1-\frac{1}{2}=\frac{1}{2}$; and the first term, 1, divided by $\frac{1}{2}$, gives 2, for the sum of all the terms.

2. What is the sum of the infinite series 1, $\frac{1}{3}$, $\frac{1}{9}$, $\frac{1}{27}$, &c.? *Ans.* $1\frac{1}{2}$.

3 What is the sum of the infinite series $\frac{1}{10}$, $\frac{3}{100}$, $\frac{9}{1000}$, $\frac{27}{10000}$, &c.? *Ans.* $\frac{1}{7}$.

4. What is the sum of the infinite series $\frac{2}{10}$, $\frac{4}{100}$, $\frac{8}{1000}$, $\frac{16}{10000}$, &c.? *Ans.* $\frac{1}{4}$.

5. What is the sum of the infinite series $\frac{1}{100}$, $\frac{3}{10000}$, $\frac{9}{1000000}$, &c.? *Ans.* $\frac{1}{97}$.

6. What is the sum of the infinite series $\frac{1}{10}$, $\frac{6}{100}$, $\frac{36}{1000}$, $\frac{216}{10000}$, &c ? *Ans.* $\frac{1}{4}$.

CHAPTER XII.

MISCELLANEOUS QUESTIONS SOLVED BY ANALYSIS.

84. A MAN and his wife usually drank out a cask of beer in 12 days; but when the man was from home, it lasted the woman 30 days. How many days would the man alone be in drinking it?

Solution.

Since it requires 12 days for the man and his wife to drink out the cask, they must, in each day, drink $\frac{1}{12}$ of it.

Again, since the woman is 30 days in drinking it, she must, in each, day, drink $\frac{1}{30}$ of it.

Hence, the fractional part which the man drank in 1 day, must be $\frac{1}{12}-\frac{1}{30}=\frac{1}{20}$; $\therefore$ in 20 days he could drink the whole.

2. A person bought several gallons of wine for $94, and after using 7 gallons himself, sold $\frac{1}{4}$ of the remainder for $20. How many gallons had he at first?

Solution.

Since he sold $\frac{1}{4}$ of the remainder, after using 7 gallons, for $20, he could have sold the whole of the remainder for $80; therefore, the value of the 7 gallons which he used was $94-80=\$14$; and 1 gallon must have cost $\frac{14}{7}=\$2$. The wine being worth $2 per gallon, he must have purchased $\frac{94}{2}=47$ gallons.

3. A person in play lost $\frac{1}{4}$ of his money, and then won 3 shillings, after which he lost $\frac{1}{3}$ of what he then had; and this done, he found that he had but 12 shillings remaining. How much had he at first?

Solution.

This, like all other questions of a similar nature, is most readily solved by an inverse operation. Thus, since the 12 shillings which finally remained, was only $\frac{2}{3}$ of what he had before losing $\frac{1}{3}$, we may find what he had before losing the $\frac{1}{3}$, by dividing 12 shillings by $\frac{2}{3}$, which thus becomes $\frac{3}{2}$ of 12 shillings$=$18 shillings; now, deducting the 3 shillings which he won, we get 15 shillings, which is $\frac{3}{4}$ of what he had at first; therefore, $\frac{4}{3}$ of 15 shillings$=$20 shillings, is what he had at first.

4. A fish was caught whose tail weighed 9 pounds; his head weighed as much as his tail and half his body, and his body weighed as much as his head and tail together. What was the weight of the fish?

Solution.

Since the head of the fish is equal to $\frac{1}{2}$ of the body, together with the tail, which equals 9 lbs., it follows that the head and tail together must equal $\frac{1}{2}$ of the body$+$18 lbs. But, by the question, the head and tail together is equal to the whole body; $\therefore$ we have this relation: $\frac{1}{2}$ of the body$+$18 pounds, must equal the whole body; consequently, $\frac{1}{2}$ of the body must equal 18 pounds, and the whole body is 36 pounds. And, since the body weighed as much as the head and tail together, it follows that the weight of the whole fish was twice that of the body, or 8 times that of the tail, which is 72 pounds.

5. A person engaged a workman for 48 days. For each day that he labored he received 24 cents, and for each day that he was idle, he paid 12 cents for his board. At the end of the 48 days, the account was settled, when the laborer received \$5·04. Required the number of working days, and the number of days he was idle.

Solution.

Had he worked all the time, he would have received $24 \times 48 = \$11·52$; but he received only \$5·04. Therefore, by being idle, he lost $\$11·52 - \$5·04 = \$6·48$. Now, for each idle day, he loses the 24 cents which he might have earned, as well as the 12 cents which he gives for his board; so that every idle day is to him a loss of $24 + 12 = 36$ cents. But we have just shown that his total loss was \$6·48; ∴ the number of idle days was $\frac{648}{36} = 18$, and he worked $48 - 18 = 30$ days.

6. A gentleman bought two pieces of silk, which, together, measured 36 yards. Each of them cost as many shillings per yard as there were yards in the piece, and their whole prices were as 4 to 1. What were the lengths of the pieces?

Solution.

Since each piece cost per yard as many shillings as there were yards in its length, it follows that their values, expressed in shillings, must be as the squares of their lengths. By the question, their prices were as 4 to 1; therefore, the squares of their lengths must be to each other as 4 to 1; consequently, their lengths must be to each other as 2 to 1.

The question is now reduced to the following: divide

36 into 2 parts, which shall be to each other as 2 to 1. These parts are $\frac{2}{3}$ of $36=24$, and $\frac{1}{3}$ of $36=12$.

7. In a mixture of wine and cider, $\frac{1}{2}$ of the whole, $+25$ gallons, is wine, and $\frac{1}{3}$ of the whole, -5 gallons, is cider. How many gallons were there of each?

Solution.

By the question, the wine$=\frac{1}{2}$ of the whole$+25$ gallons, and the cider$=\frac{1}{3}$ of the whole-5 gallons. Hence, taking the sum of these expressions, we get the whole $=(\frac{1}{2}+\frac{1}{3})$ or $\frac{5}{6}$ of the whole $+$ 20 gallons; $\therefore$ $\frac{1}{6}$ of the whole equals 20 gallons; consequently, the whole is 120 gallons.

Now $\frac{1}{2}$ of the whole is 60 gallons, to which, add 25 gallons, we get, for the wine, 85 gallons.

Again, $\frac{1}{3}$ of the whole is 40 gallons, from which, subtracting 5 gallons, we get, for the cider, 35 gallons.

8. A market-woman bought a certain number of eggs, at 2 for a penny, and as many more at 3 for a penny; and, having sold them again, all together, at the rate of 5 for 2 pence, found that she had lost 4 pence. How many eggs had she?

Solution.

Since, by the question, half of the eggs cost $\frac{1}{2}$ of a penny apiece, and the other half cost $\frac{1}{3}$ of a penny apiece, it follows that the average price which she gave for the eggs was $(\frac{1}{2}+\frac{1}{3})\div 2=\frac{5}{12}$ of a penny apiece. Since she sold them all together at the rate of 5 for 2 pence, that is, $\frac{2}{5}$ of a penny apiece, she must have lost on each egg $\frac{5}{12}-\frac{2}{5}=\frac{1}{60}$ of a penny. Therefore, to lose 1 penny, she must have disposed of 60 eggs, and to lose 4 pence, she must have had 240 eggs.

9. A and B can, together, do a piece of work in 8 days; A and C can, together, do it in 9 days; B and C can, together, do it in 10 days. How many days would it require for each to perform the work alone?

Solution.

Since A and B can do the work in 8 days, they can, in 1 day, do $\frac{1}{8}$ part of it; for a similar reason, A and C can do $\frac{1}{9}$ part of it in 1 day; B and C can do $\frac{1}{10}$ part of it in 1 day. Adding these fractional parts together, and observing that each individual has been included twice, we shall get, by dividing the sum by 2, the following fraction, $(\frac{1}{8}+\frac{1}{9}+\frac{1}{10})\div 2=\frac{121}{720}$, which is the fractional part of the work which they all together would perform in 1 day.

We have already seen that the part which B and C can perform in 1 day is $\frac{1}{10}$; $\therefore$ $\frac{121}{720}-\frac{1}{10}=\frac{49}{720}$, is the fractional part which A could perform in 1 day. Hence, the time in which A could alone perform the work is $\frac{720}{49}=14\frac{34}{49}$ days.

Again, the fractional part which A and C together could perform, is $\frac{1}{9}$; $\therefore$ $\frac{121}{720}-\frac{1}{9}=\frac{41}{720}$, is the fractional part which B could perform in 1 day; hence, the time in which B could alone perform the work, is $\frac{720}{41}=17\frac{23}{41}$ days. The fractional part which A and B together could perform, is $\frac{1}{8}$; $\therefore$ $\frac{121}{720}-\frac{1}{8}=\frac{31}{720}$, is the fractional part which C could perform in 1 day; hence, the time in which he could alone perform the work is $\frac{720}{31}=23\frac{7}{31}$ days.

10. A and B have each the same income. A contracts an annual debt, amounting to $\frac{1}{7}$ of it; B lives on $\frac{4}{5}$ of it; and, at the end of ten years, B lends to A enough to pay off his debts, and has \$160 left. What is the income?

Solution.

Since B lives on $\frac{4}{5}$ of his income, he must save $\frac{1}{5}$ of it. A's debt for 1 year being $\frac{1}{7}$ of the income, B will have left, after paying A's debt, $\frac{1}{5}-\frac{1}{7}=\frac{2}{35}$ of his income. And, since this would in 10 years amount to \$160, $\frac{2}{35}$ of his income must equal \$16. Hence, the income was $\frac{35}{2}$ of $\$16=\$280$.

11. A merchant supported himself 3 years for \$50 a year; at the end of each year, he added to that part of his stock which was not thus expended, a sum equal to $\frac{1}{3}$ of this part. At the end of the third year, his original stock was doubled. What was the stock?

Solution.

After supporting himself the first year, he will have his original stock, $-\$50$; this, increased by its third part, will become $\frac{4}{3}$ of his original stock, $-\frac{4}{3}$ of \$50; living upon another \$50, he will have left $\frac{4}{3}$ of his original stock, $-\frac{4}{3}$ of $\$50-\$50$. This must again be increased by its third part, giving $\frac{16}{9}$ of his original stock, $-\frac{16}{9}$ of $\$50-\frac{4}{3}$ of \$50.

Again, living upon \$50, he will have left $\frac{16}{9}$ of his original stock, $-\frac{16}{9}$ of $\$50-\frac{4}{3}$ of $\$50-\$50$; increasing this once more by its third part, we get $\frac{64}{27}$ of his original stock, $-\frac{64}{27}$ of $\$50-\frac{16}{9}$ of $\$50-\frac{4}{3}$ of \$50. This, by the question, is equal to twice his original stock, or to $\frac{54}{27}$ of his original stock.

Hence, $\frac{64}{27}-\frac{54}{27}=\frac{10}{27}$ of his original stock must equal $\frac{64}{27}$ of $\$50+\frac{16}{9}$ of $\$50+\frac{4}{3}$ of $\$50=\frac{148}{27}$ of $\$50=\frac{7400}{27}$ dollars; $\therefore$ his stock was $\frac{7400}{27}\div\frac{10}{27}=\$740$.

12. Fourteen oxen have, in 3 weeks, eaten all the

grass which grew on 2 acres of land, in such a manner, that they not only ate all the grass which at first was there, but also that which grew during the time they were grazing. In like manner have 16 oxen, in 4 weeks, eaten all the grass upon 3 acres of land. How many oxen can, in this way, graze for 5 weeks upon 6 acres of land?

Solution.

	Acres.	Oxen.	Weeks	
(1.)	2	14	3	By 1st condition.
(2.)	1	7	3	
(3.)	3	7	9	
(4.)	3	63	1	
(5.)	3	16	4	By 2d condition.
(6.)	3	64	1	

If the grass of 2 acres, with its growth for 3 wks., keep 14 oxen for 3 weeks, then will the grass of one acre, with its growth for 3 weeks, keep 7 oxen for 3 weeks; and the grass of 3 acres, with its growth for 3 weeks, will keep 7 oxen 9 weeks, or 63 oxen for 1 week, which result corresponds with (4) in the adjoining table.

Again, by the second condition of the question, if the grass of 3 acres, with its growth for 4 weeks, keep 16 oxen for 4 weeks, then will the grass of 3 acres, with its growth for 4 weeks, keep 64 oxen for 1 week, which corresponds with (6).

By carefully comparing (4) and (6), we see that the growth of 3 acres for 1 week is sufficient to keep 1 ox 1 week; consequently, the growth of 6 acres for 1 week will keep 2 oxen for 1 week; or, which is the same thing, the growth of 6 acres for 5 weeks will keep 2 oxen for 5 weeks.

By (4) we have seen that the grass of 3 acres, with its growth for 3 weeks, will keep 63 oxen for 1 week; but the growth of 3 acres for 3 weeks, will keep 3 oxen

1 week; consequently, the grass alone of 3 acres will keep 60 oxen 1 week, and the grass of 6 acres will keep 120 oxen for 1 week, or it will keep 24 oxen for 5 weeks. Hence, finally, the grass of 6 acres, with its growth for 5 weeks, will keep, during 5 weeks, 2+24=26 oxen.

13. A person expends just £100 for live stock, consisting of geese, sheep, and cows; for each goose he paid 1 s., for each sheep £1, and for each cow £5. How many did he purchase of each kind, so as to have just 100 in all?

Solution.

Since the average price of the animals was £1=20 s., the price of a goose was 19 s. below the average, and the price of a cow was 80 s. above the average.

Hence, were he to purchase the geese and cows only in the ratio of 80 geese to 19 cows, their whole cost would be just as many pounds sterling as there were animals; thus, by purchasing 80 geese and 19 cows, he would have 99 animals, together worth £99. Now, by adding 1 sheep, worth £1, he will have 100 animals, together worth £100. So that he bought 80 geese, 1 sheep, and 19 cows.

14. A and B leave Utica for Albany at the same time that C leaves Albany for Utica. If A goes 8 miles each hour, B 12 miles, and C 9, when will C be equally distant between A and B, if the distance between Utica and Albany is 95 miles?

Solution.

The average velocity of A and B is 10 miles an hour; hence, if D, a fourth person, leave Utica at the same time, going 10 miles each hour, he will always be equally distant between A and B, and the time sought will be when he is met by C. Now D and C together travel 19 miles each hour; consequently, they will meet in $95 \div 19 = 5$ hours, at which time C will be equally distant from A and B.

15 A, B, C, D, and E, play together on this condition: that he who loses shall give to all the rest as much as they already have. First, A loses, then B, then C, then D, and at last also E. All lose in turn, and yet, at the end of the fifth game, they have all the same sum, viz., each \$32. How much had each before they began to play?

Solution.

The solution of this question is the most readily effected by a reverse process; that is, by beginning with the last game, and playing them all in a reverse order, as follows: first, take from A, B, C, and D, half they have, and add it to E's money; second, take from A, B, C, and E, half what they now have, and add it to D's; third, take from A, B, D, and E, half what they now have, and add it to C's; fourth, take half of A's, C's, D's, and E's, and add it to B's; lastly, take half of B's, C's, D's, and E's, and add it to A's.

These successive operations may be exhibited as follows:

	A.	B.	C.	D.	E.	
	\$32	\$32	\$32	\$32	\$32,	end of 5th game,
	16	16	16	16	96,	end of 4th game,
	8	8	8	88	48,	end of 3d game,
	4	4	84	44	24,	end of 2d game,
	2	82	42	22	12,	end of 1st game,
Ans.	81	41	21	11	6,	before playing.

16. A father left to his three sons, whose ages are 8, 10, and 13 years, \$10000, to be so divided that the respective parts, being placed out at 5 per cent., compound interest, should amount to equal sums when they became 21 years of age. What are the parts?

Solution.

By the question, their respective shares would be at interest, 13, 11, and 8 years.

We find, by table under Art. **69,** the present worth of \$1 for 13, 11, and 8 years respectively, at 5 per cent., compound interest, to be \$0·530321, \$0·584679, and \$0·676839. Now it is obvious that the parts must be to each other in the same ratio as the numbers 530321, 584679, and 676839; the sum of these numbers is 1791839. Hence, the parts are as follows:

The 1st one's part is $\frac{530321}{1791839}$ of \$10000=\$2959·646.
The 2d one's part is $\frac{584679}{1791839}$ of \$10000=\$3263·011.
The 3d one's part is $\frac{676839}{1791839}$ of \$10000=\$3777·343.

17. Find what each of the four persons, A, B, C, and D, are worth, by knowing,

1st. That A's money, together with $\frac{1}{3}$ of B's, C's, and D's, is equal to \$137.

2d. That B's money, together with $\frac{1}{4}$ of A's, C's, and D's, is equal to \$137.

3d. That C's money, together with $\frac{1}{5}$ of A's, B's, and D's, is equal to \$137.

4th. That D's money, together with $\frac{1}{6}$ of A's, B's, and C's, is equal to \$137.

Solution.

It is evident that $\frac{1}{3}$ of B's, C's, and D's money is the same as $\frac{1}{3}$ of the *sum of all*, MINUS $\frac{1}{3}$ of A's; therefore,

A's, together with $\frac{1}{3}$ of B's, C's, and D's, is equal to A's$+\frac{1}{3}$ of the *sum of all*$-\frac{1}{3}$ of A's, which, by the first condition, equals \$137.

Consequently, $\frac{2}{3}$ of A's$=\$137-\frac{1}{3}$ of the *sum of all*. $\therefore$ A's$=\frac{3}{2}$ of $\$137-\frac{1}{2}$ of the *sum of all*. In a similar way, we get B's$=\frac{4}{3}$ of $\$137-\frac{1}{3}$ of the *sum of all*.

C's$=\frac{5}{4}$ of $\$137-\frac{1}{4}$ of the *sum of all*.

D's$=\frac{6}{5}$ of $\$137-\frac{1}{5}$ of the *sum of all*.

Taking the sum of these values of A, B, C, and D, we get the *sum of all*$=(\frac{3}{2}+\frac{4}{3}+\frac{5}{4}+\frac{6}{5})$ of $\$137-(\frac{1}{2}+\frac{1}{3}+\frac{1}{4}+\frac{1}{5})$ of the *sum of all*. $\therefore (1+\frac{1}{2}+\frac{1}{3}+\frac{1}{4}+\frac{1}{5})$ of the *sum of all*$=(\frac{3}{2}+\frac{4}{3}+\frac{5}{4}+\frac{6}{5})$ of \$137. And the *sum of all*$=\dfrac{\frac{3}{2}+\frac{4}{3}+\frac{5}{4}+\frac{6}{5}}{1+\frac{1}{2}+\frac{1}{3}+\frac{1}{4}+\frac{1}{5}}$ of $\$137=\frac{317}{137}$ of $\$137=\$317$.

This value for the *sum of all*, being substituted in the above values of A, B, C, and D, we obtain the following results:

A's$=\frac{3}{2}$ of $\$137-\frac{1}{2}$ of $\$317=\$\ 47$.
B's$=\frac{4}{3}$ of $\$137-\frac{1}{3}$ of $\$317=\$\ 77$.
C's$=\frac{5}{4}$ of $\$137-\frac{1}{4}$ of $\$317=\$\ 92$.
D's$=\frac{6}{5}$ of $\$137-\frac{1}{5}$ of $\$317=\$101$.

NOTE.—It is obvious that this method of solving the above question will apply in the case of any number of unknown quantities which are similarly related to each other.

18. A and B, settling accounts, found that if £6 were added to $\frac{2}{3}$ of A's money, and the same sum taken from $\frac{2}{3}$ of B's, the *sum* would be $\frac{7}{8}$ of the *remainder*, and that the sum and the remainder, added together, made £72. What was each man's money?

Solution.

Since £6 was added to $\frac{2}{3}$ of A's money, and subtracted from $\frac{2}{3}$ of B's money, the sum of A's and B's money, after this addition and subtraction, is the same as it would have been had no such addition and subtraction been made. Therefore, $\frac{2}{3}$ of A's and B's money is, by the question, £72.

Again, by the question, $\frac{2}{3}$ of A's, increased by £6, is equal to $\frac{7}{8}$ of $\frac{2}{3}$ of B's, diminished by £6; $\therefore$ $\frac{2}{3}$ of A's, increased by £6, is to $\frac{2}{3}$ of B's, diminished by £6, as 7 to 8. But we have already seen that $\frac{2}{3}$ of A's, increased by £6, added to $\frac{2}{3}$ of B's, diminished by £6, is £72. Hence, if we divide £72 into two parts which are to each other as 7 to 8, these parts will be $\frac{2}{3}$ of A's, increased by £6, and $\frac{2}{3}$ of B's, diminished by £6. The parts of £72 which are to each other as 7 to 8, are $\frac{7}{15}$ of £72=£33$\frac{3}{5}$, and $\frac{8}{15}$ of £72=£38$\frac{2}{5}$. Therefore, $\frac{2}{3}$ of A's, increased by £6, is equal to £33$\frac{3}{5}$; consequently, $\frac{2}{3}$ of A's is £33$\frac{3}{5}$−£6=£27$\frac{3}{5}$; and the whole of A's money is $\frac{3}{2}$ of £27$\frac{3}{5}$=£41$\frac{2}{5}$. And $\frac{2}{3}$ of B's, diminished by £6, is equal to £38$\frac{2}{5}$. Therefore, $\frac{2}{3}$ of B's is £38$\frac{2}{5}$ +£6=£44$\frac{2}{5}$; and the whole of B's money is $\frac{3}{2}$ of £44$\frac{2}{5}$=£66$\frac{3}{5}$.

19. A purse of \$2850 is to be divided among three persons, A, B, and C. A's share is to be to B's as 6 to 11, and C is to have \$300 more than A and B together. What is each one's share?

Solution.

Since C is to have \$300 more than A and B together, it follows that A and B must have half of what is left, after subtracting \$300; hence, A and B together have \$1275; this, divided into two parts, which are to each other as 6 to 11, gives A's=$\frac{6}{17}$ of \$1275=\$450; B's=$\frac{11}{17}$ of \$1275=\$825; C's is evidently \$1575.

20. Two persons, A and B, purchase in company 100 acres of land for \$1000, of which the southern portion is of rather the best quality. In the division of it, A says to B, let me have the southern portion, and I will pay for my part $\$1\frac{1}{18}$ per acre more than you pay for yours. How much land must each have, and at what price per acre?

NOTE.—This, and like questions, can be solved by the following

RULE.

Divide half the whole cost by the whole number of acres, and to the square of the quotient add the square of half the difference of the prices per acre; then extract the square root of the sum, and to this root add the quotient of half the whole cost, divided by the whole number of acres. This last sum, increased by half the difference of the prices per acre, will give the price per acre of the best land; and, diminished by the same, will give the price per acre of the poorest.

Applying the above rule, we find the quotient of half the whole cost, divided by the whole number of acres, to be 5, which, squared, gives 25; this, increased by the square of half the difference of the price per acre, becomes $25+\frac{361}{1296}=\frac{32761}{1296}$, whose square root is $\frac{181}{36}$; this root, added to 5, gives $\frac{361}{36}=10\frac{1}{36}$. Therefore, the price per acre of A's land, is $10\frac{1}{36}+\frac{19}{36}=\$10\frac{5}{9}$. The price of B's land is $10\frac{1}{36}-\frac{19}{36}=\$9\frac{1}{2}$.

$\$500\div\$10\frac{5}{9}=47\frac{7}{19}$ acres, for A's portion.

$\$500\div\$\ 9\frac{1}{2}=52\frac{12}{19}$ acres, for B's portion.

21. A boy divided his apples among his four companions in the following manner: To the first he gave half an apple more than half his whole number; to the second he likewise gave half an apple more than half the number which he then had; in the same manner he

divided with the third and fourth companion, giving to each half an apple more than half the number which was left after giving to the preceding one. After having divided with the fourth companion, he had but one apple left. How many had he at first?

Solution.

This question is most easily solved by beginning with the last companion, and reversing each successive operation.

Since he had but one apple left after making the last division, he must have had 3 after the third division, because 3, diminished by $\frac{1}{2}$ more than its half, leaves 1; for a similar reason, he must have had 7 after the second division. In this way we may retrace the process, by multiplying the number by 2, and adding 1 to the product for each successive step. In this way we find that, after the first division, he must have had $7\times2+1=15$. And, before he divided with the first, he must have had $15\times2+1=31$ apples.

NOTE.—From the above method of solving this question, we see that it would not be difficult to extend the solution to the case of any number of divisions. Indeed, we see that the number required is always one less than 2, raised to a power whose index is one more than the number of divisions.

Thus, had there been 10 companions to divide with, after the above method, he must have had $2^{11}-1=2047$ apples at first.

This method of division is effected without dividing an individual apple.

22. A hare is 50 leaps before a greyhound, and takes 4 leaps to the greyhound's 3, but 2 leaps of the hound are equal to 3 of the hare's. How many leaps must the greyhound take before he catches the hare?

Solution.

Since 2 leaps of the greyhound equals 3 leaps of the hare, it follows that 6 leaps of the greyhound equals 9 leaps of the hare.

But, while the greyhound takes 6 leaps, the hare takes 8 leaps; therefore, while the hare takes 8 leaps, the greyhound gains on her 1 leap. Hence, to gain 50 leaps, she must take $50 \times 8 = 400$ leaps; but, while the hare takes 400 leaps, the greyhound would take 300 leaps, since the number of leaps taken by them were as 4 to 3.

23. From a cask of wine a tenth part is drawn out, and then the cask is filled with water; after which, a tenth part of the mixture is drawn out; again the cask is filled, and again a tenth part of the mixture is drawn out. Now suppose the fluids mix uniformly at each time the cask is replenished, what fractional part of wine will remain after the process of drawing out and replenishing has been repeated ten times?

Solution.

Since $\frac{1}{10}$ of the wine is drawn out at the first drawing, there must remain $\frac{9}{10}$. And after the cask is filled with water, $\frac{1}{10}$ of the whole being drawn out, there will remain $\frac{9}{10}$ of the mixture; but $\frac{9}{10}$ of the mixture, we have already seen, is wine; therefore, after the second drawing, there will remain $\frac{9}{10}$ of $\frac{9}{10}$ of wine, or $\frac{9^2}{10^2}$. By similar reasoning, we see that after the third drawing there will remain $\frac{9}{10}$ of $\frac{9}{10}$ of $\frac{9}{10}$ of wine, or $\frac{9^3}{10^3}$.

From this, we see that the part of wine remaining is expressed by the ratio $\frac{9}{10}$, raised to a power whose exponent is the number of times the cask has been drawn from. Hence, in the present question, the fractional part of wine is $\frac{9^{10}}{10^{10}} = \frac{3486784401}{10000000000} = 0{\cdot}3486784401$, which is nearly 35 per cent.

24. There are two vessels of equal capacity, the one $\frac{2}{3}$ full of wine, the other $\frac{2}{3}$ full of water; now suppose the vessel containing the wine to be first filled from the water vessel, and then the water vessel to be filled from the wine vessel; and so on, alternately filling the one from the other. Now, on the supposition that the fluids mix uniformly at each time, how much wine, and how much water will the wine vessel contain after it has been filled 5 times? And how much water, and how much wine will the water vessel contain after it has been filled 5 times?

Solution.

In the first place, the wine vessel is filled from the water, so that it will consist of $\frac{2}{3}$ of wine, and $\frac{1}{3}$ of water, and there will remain in the water vessel $\frac{1}{3}$ of water.

Now the water vessel is filled from the wine vessel by drawing from it $\frac{2}{3}$; that is, $\frac{2}{3}$ of $\frac{2}{3}$ of wine $= \frac{4}{9}$ of wine, and $\frac{2}{3}$ of $\frac{1}{3}$ of water $= \frac{2}{9}$ of water, so that the water vessel consists of $\frac{5}{9}$ of water and $\frac{4}{9}$ of wine; and there remains in the wine vessel $\frac{2}{9}$ of wine, and $\frac{1}{9}$ of water.

We next fill the wine vessel, and find that it will contain $\frac{14}{27}$ of wine, and $\frac{13}{27}$ of water, and that there will remain in the water vessel $\frac{5}{27}$ of water, and $\frac{4}{27}$ of wine.

These successive operations will be more clearly comprehended by the aid of the following

TABLE.

	WATER.	WINE.	WATER.	WINE.	
	$\frac{2}{3}$			$\frac{2}{3}$	
	$\frac{1}{3}$		$\frac{1}{3}$	$\frac{2}{3}$	Wine vessel filled 1st time.
Water vessel filled 1st time.	$\frac{5}{9}$	$\frac{4}{9}$	$\frac{1}{9}$	$\frac{2}{9}$	
	$\frac{5}{27}$	$\frac{4}{27}$	$\frac{13}{27}$	$\frac{14}{27}$	Wine vessel filled 2d time.
Water vessel filled 2d time.	$\frac{41}{81}$	$\frac{40}{81}$	$\frac{13}{81}$	$\frac{14}{81}$	
	$\frac{41}{243}$	$\frac{40}{243}$	$\frac{121}{243}$	$\frac{122}{243}$	Wine vessel filled 3d time.
Water vessel filled 3d time.	$\frac{365}{729}$	$\frac{364}{729}$	$\frac{121}{729}$	$\frac{122}{729}$	
	$\frac{365}{2187}$	$\frac{364}{2187}$	$\frac{1093}{2187}$	$\frac{1094}{2187}$	Wine vessel filled 4th time.
Water vessel filled 4th time.	$\frac{3281}{6561}$	$\frac{3280}{6561}$	$\frac{1093}{6561}$	$\frac{1094}{6561}$	
	$\frac{3281}{19683}$	$\frac{3280}{19683}$	$\frac{9841}{19683}$	$\frac{9842}{19683}$	Wine vessel filled 5th time.
Water vessel filled 5th time.	$\frac{29525}{59049}$	$\frac{29524}{59049}$	$\frac{9841}{59049}$	$\frac{9842}{59049}$	

The law of continuation is so simple, that this table might with ease be carried to any extent desired. The denominators of the partial fractions expressing the wine and water, are the successive powers of 3, and the numerators are found by dividing the denominators into two parts, differing from each other by a unit. Thus, after the 5th pouring, the wine vessel has been filled 3 times, and the quantity of wine is $\frac{\frac{1}{2}(3^5+1)}{3^5}$ $=\frac{122}{243}$, and the water is $\frac{\frac{1}{2}(3^5-1)}{3^5}=\frac{121}{243}$. After the 6th pouring, the water vessel has been filled 3 times, and the water is $\frac{\frac{1}{2}(3^6+1)}{3^6}=\frac{365}{729}$, and the wine is $\frac{\frac{1}{2}(3^6-1)}{3^6}$ $=\frac{364}{729}$.

We may remark that an odd number of pourings leaves the wine vessel full, and, consequently, the water vessel $\frac{1}{3}$ full; and an even number of pourings leaves the water vessel full, and, consequently, the wine vessel $\frac{1}{3}$ full.

We may also remark that the fluids will never become equally mixed, since the denominators which express the fractional parts of wine and water, at any particular time, are always the same, while their numerators differ a unit; so that the wine vessel will always contain an excess of wine, and the water vessel an excess of water. Notwithstanding they are never equal, still they approach more nearly to an equality the greater the number of times the operation is performed.

25. A person possesses a wagon with a mechanical contrivance, by which the difference of the number of revolutions of the wheels on a journey may be determined. It is known that each of the fore wheels are $5\frac{1}{4}$ feet, and that each of the hind wheels are $7\frac{1}{8}$ feet in

circumference. Now, when on a journey, the fore wheel has made 2000 revolutions more than the hind wheel; how great was the distance traveled?

Solution.

Circumference of fore wheel $=5\frac{1}{4}$ ft. $=\frac{21}{4}=\frac{42}{8}$ feet.
" hind wheel $=7\frac{1}{8}$ ft. $=\frac{57}{8}$ feet.

Now find the least number that will divide by 42 and 57, (by Rules under Art. **10** or **11.**) This number is 798. Hence, the least number which will divide by $\frac{42}{8}$ and $\frac{57}{8}$ is $\frac{798}{8}$.

Now, in going $\frac{798}{8}$ feet, the fore wheel turns over $\frac{798}{8}\div\frac{42}{8}=19$ times; and the hind wheel turns over $\frac{798}{8}\div\frac{57}{8}=14$ times; therefore, in making a journey of $\frac{798}{8}$ feet, the fore wheel turns over $19-14=5$ times more than the hind wheel. Hence, in order that the fore wheel may turn over 2000 times more than the hind wheel, he must perform a journey of $\frac{798}{8}\times\frac{2000}{5}=39900$ feet, the answer.

26. Suppose $1 had been put out at compound interest, at 7 per cent., the 5th day of October, 1585, how much would it have amounted to on the 5th day of October, 1841?

Solution.

By referring to what has been said under the subject of compound interest, we see that we must find the 256th power of 1·07. After reaching the 4th power, the subsequent multiplications have been performed by the abridged method, Art. **39.**

1·07 = 1 year.
1·07
749
107
1·1449 = 2 yrs.
1·1449
103041
45796
45796
11449
11449
1·310796010 = 4 yrs.
1·310796010
1310796010
393238803
13107960
917557
117972
7865
13
1·718186180 = 8 yrs.
1·718186180
1718186180
1202730326
17181862
13745489
171819
137455
10309
172
137
2·952163749 = 16 yrs.

2·952163749
2·952163749
5904327498
2656947374
147608187
5904327
295216
177130
8856
2066
118
26
8·715270798 = 32 yrs.
8·715270798
69722166384
6100689559
87152708
43576354
1743054
610069
6101
784
70
75·955945083 = 64 yrs.
75·955945083
531691615581
37977972541
6836035057
379779725
37977973
6836035
303824
37978
608
23
5769·305593425 = 128yrs.

```
        5769·30559345
        5769·30559345
       ──────────────
       2884652796725
        403851391542
         34615833561
          5192375034
           173079168
             2884652
               51924
                1731
                 231
                  29
       ──────────────
Ans. $33284887·03062=256 years.
```

27. Suppose from an acorn there shoots up a single stalk, at the end of the year; that, at the end of each year thereafter, this stalk puts forth as many new branches as it is years old; also suppose all the branches to follow the same law, that is, to produce as many new branches as they are years old. How many branches will this oak tree consist of at the end of 20 years.

*Solution.**

From the conditions of the question, we know that at the end of the *first* year there will be simply one stalk, or branch, which we will denote by 1_0; at the end of the *second* year, this branch will become 1 year old, and produce a new branch, so that we shall have 1_1+1_0; at the end of the *third* year, the branches, 1_1+1_0, will become 1_2+1_1, the first of which being 2 years of age, will produce 2 new branches; the other will produce

* For an algebraic solution of this, see Recurring Series in my ALGEBRA.

1 new one; we shall, therefore, have $1_2 + 1_1 + 3_0$. Proceeding in this way, we obtain the following results:

End of	1st year,		1_0	$= 1$
"	2d	"	$1_1 + 1_0$	$= 2$
"	3d	"	$1_2 + 1_1 + 3_0$	$= 5$
"	4th	"	$1_3 + 1_2 + 3_1 + 8_0$ . . .	$= 13$
"	5th	"	$1_4 + 1_3 + 3_2 + 8_1 + 21_0$	$= 34$
"	6th	"	$1_5 + 1_4 + 3_3 + 8_2 + 21_1 + 55_0$	$= 89$

In this scheme, the small figures at the bottom of the larger ones, denote the age, in years, of the branches to which they are attached. Thus, at the end of the fifth year, there will be 1 branch 4 years old, 1 branch 3 years old, 3 branches 2 years old, 8 branches 1 year old, and 21 new branches of no age.

The law of the above series is obvious. It is such, *that twice any term, increased by the sum of all the preceding terms, gives the next succeeding term. Or, three times any term, decreased by the next preceding term, will give the next succeeding term.*

These terms may be found most easily by continual addition, as given by the following table, where each succeeding term is found by adding the two preceding ones.

		0	New branches	1st year.
Total branches	1st year, . . .	1		
		1	New branches	2d year.
Total branches	2d year,	2		
		3	New branches	3d year.
Total branches	3d year, . . .	5		
		8	New branches	4th year.
Total branches	4th year, . . .	13		
		21	New branches	5th year.
Total branches	5th year, . . .	34		
		55	New branches	6th year.
Total branches	6th year, . . .	89		

	144	New branches 7th year.
Total branches 7th year, . .	233	
	377	New branches 8th year.
Total branches 8th year. . .	610	
	987	New branches 9th year.
Total branches 9th year, . .	1597	
	2584	New branches 10th year.
Total branches 10th year, . .	4181	
	6765	New branches 11th year.
Total branches 11th year, .	10946	
	17711	New branches 12th year.
Total branches 12th year, .	28657	
	46368	New branches 13th year.
Total branches 13th year, .	75025	
	121393	New branches 14th year.
Total branches 14th year, .	196418	
	317811	New branches 15th year.
Total branches 15th year, .	514229	
	832040	New branches 16th year.
Total branches 16th year, .	1346269	
	2178309	New branches 17th year.
Total branches 17th year, .	3524578	
	5702887	New branches 18th year.
Total branches 18th year, .	9227465	
	14930352	New branches 19th year.
Total branches 19th year,	24157817	
	39088169	New branches 20th year.
Total branches 20th year,	63245986	

28. Find ten numbers, such that the first, with $\frac{1}{5}$ of all the others, shall make 845693; the second, with $\frac{1}{6}$ of all the others, shall make 845693; the third, with $\frac{1}{7}$ of all the others, shall make 845693; the fourth with $\frac{1}{8}$, the fifth with $\frac{1}{9}$, the sixth with $\frac{1}{10}$, the seventh with $\frac{1}{11}$, the eighth with $\frac{1}{12}$, the ninth with $\frac{1}{13}$, and the tenth with $\frac{1}{14}$ of all the others, shall make, respectively, the same number, 845693.

Solution.

This may be solved by a method analogous to the one employed for the solution of question 17, pages 257–8.

Thus, $\frac{1}{5}$ of the 2d, 3d, 4th, 5th, 6th, 7th, 8th, 9th, and 10th, is the same as $\frac{1}{5}$ of the *sum of all*, MINUS $\frac{1}{5}$ of the 1st. Therefore, the 1st, together with $\frac{1}{5}$ of the others, is equal to the 1st $+\frac{1}{5}$ of the *sum of all* $-\frac{1}{5}$ of the 1st, which, by the first condition of the question, equals 845693. Consequently, $\frac{4}{5}$ of the 1st$=845693-\frac{1}{5}$ of the *sum of all.*

Therefore, the 1st$=\frac{5}{4}$ of $845693-\frac{1}{4}$ of the *sum of all.*

In a similar way, we get,

2d $=\frac{6}{5}$ of $845693-\frac{1}{5}$ of the *sum of all.*
3d $=\frac{7}{6}$ of $845693-\frac{1}{6}$ of the *sum of all.*
4th$=\frac{8}{7}$ of $845693-\frac{1}{7}$ of the *sum of all.*
5th$=\frac{9}{8}$ of $845693-\frac{1}{8}$ of the *sum of all.*
6th$=\frac{10}{9}$ of $845693-\frac{1}{9}$ of the *sum of all*
7th$=\frac{11}{10}$ of $845693-\frac{1}{10}$ of the *sum of all.*
8th$=\frac{12}{11}$ of $845693-\frac{1}{11}$ of the *sum of all.*
9th$=\frac{13}{12}$ of $845693-\frac{1}{12}$ of the *sum of all.*
10th$=\frac{14}{13}$ of $845693-\frac{1}{13}$ of the *sum of all.*

If we take the sum of the above values, and observe that

$\frac{5}{4}+\frac{6}{5}+\frac{7}{6}+\frac{8}{7}+\frac{9}{8}+\frac{10}{9}+\frac{11}{10}+\frac{12}{11}+\frac{13}{12}+\frac{14}{13}=\frac{4088933}{360360}$,

$\frac{1}{4}+\frac{1}{5}+\frac{1}{6}+\frac{1}{7}+\frac{1}{8}+\frac{1}{9}+\frac{1}{10}+\frac{1}{11}+\frac{1}{12}+\frac{1}{13}=\frac{485333}{360360}$,

we shall obtain,

Sum of all$=\frac{4088933}{360360}$ of $845693-\frac{485333}{360360}$ of the *sum of all.*

Therefore, $(1+\frac{485333}{360360})$ of the *sum of all*$=\frac{4088933}{360360}$ of 845693, or $\frac{845693}{360360}$ of the *sum of all*$=\frac{4088933}{360360}$ of 845693.

Consequently, the *sum of all*$=4088933$.

Having found the *sum of all*, we can readily find the

respective numbers by substituting it in the foregoing expressions, thus:

1st $= \frac{5}{4}$ of 845693 $- \frac{1}{4}$ of 4088933 $=$ 34883.
2d $= \frac{6}{5}$ of 845693 $- \frac{1}{5}$ of 4088933 $=$ 197045.
3d $= \frac{7}{6}$ of 845693 $- \frac{1}{6}$ of 4088933 $=$ 305153.
4th $= \frac{8}{7}$ of 845693 $- \frac{1}{7}$ of 4088933 $=$ 382373.
5th $= \frac{9}{8}$ of 845693 $- \frac{1}{8}$ of 4088933 $=$ 440288.
6th $= \frac{10}{9}$ of 845693 $- \frac{1}{9}$ of 4088933 $=$ 485333.
7th $= \frac{11}{10}$ of 845693 $- \frac{1}{10}$ of 4088933 $=$ 521369.
8th $= \frac{12}{11}$ of 845693 $- \frac{1}{11}$ of 4088933 $=$ 550853.
9th $= \frac{13}{12}$ of 845693 $- \frac{1}{12}$ of 4088933 $=$ 575423.
10th $= \frac{14}{13}$ of 845693 $- \frac{1}{13}$ of 4088933 $=$ 596213.

If, instead of the constant number 845693, it were required to have the respective numbers, when increased by the above fractional parts of the others, equal to any other constant number, we must multiply these answers by the ratio of this new number to 845693.

[For a complete algebraic solution of this, and all other analogous questions, see my ALGEBRA.]

CHAPTER XIII.

MISCELLANEOUS QUESTIONS.

85. WHAT are the prime factors of 2006?

Ans. $2006 = 2 \times 17 \times 59$.

2. What are the prime factors of 3742?

Ans. $3742 = 2 \times 1871$.

3. What is the greatest common measure of 720, 360, and 180? *Ans.* $2^2 \times 3^2 \times 5 = 180$.

4. What is the greatest common measure of 420, 147, and 210? *Ans.* $3 \times 7 = 21$.

5. What is the least common multiple of 4, 16, 24, and 40? *Ans.* $2^4 \times 3 \times 5 = 240$.

6. What is the least common multiple of 8, 36, and 100? *Ans.* $2^3 \times 3^2 \times 5^2 = 1800$.

7. What are all the divisors of 376?

Ans. 1, 2, 4, 8, 47, 94, 188, 376.

8. What are all the divisors of 23456?

Ans. { 1, 2, 4, 8, 16, 32, 733, 1466, 2932, 5864, 11728, 23456.

9. What is the number of the divisors of 7866?

Ans. $2 \times 3 \times 2 \times 2 = 24$.

10. What is the number of the divisors of 1000?

Ans. $4 \times 4 = 16$.

11. Reduce $\frac{378}{456}$ to its lowest terms. *Ans.* $\frac{63}{76}$.

12. Reduce $\frac{123456}{234568}$ to its lowest terms.

Ans. $\frac{15432}{29321}$.

13. Reduce the improper fraction $\frac{1236}{63}$ to a mixed fraction. *Ans.* $19\frac{13}{21}$.

14. Reduce $\frac{4567}{756}$ to a mixed fraction. *Ans.* $6\frac{31}{756}$.

15. Reduce $67\frac{5}{11}$ to an improper fraction. *Ans.* $\frac{742}{11}$.

16. Reduce $37\frac{1}{3}$ to an improper fraction. *Ans.* $\frac{112}{3}$.

17. What is the product of $\frac{365}{21}$ into $\frac{35}{17}$? *Ans.* $35\frac{40}{51}$.

18. What is the product of $\frac{29}{36}$ into $\frac{108}{58}$? *Ans.* $\frac{1}{2}$.

19. Reduce the compound fraction $\frac{1}{2}$ of $\frac{2}{3}$ of $\frac{3}{7}$ of $\frac{14}{20}$ to a simple fraction. *Ans.* $\frac{1}{10}$.

20. Reduce $\frac{3}{11}$, $\frac{4}{33}$, and $\frac{6}{7}$, to fractions having a common denominator. *Ans.* $\frac{63}{231}$, $\frac{28}{231}$, $\frac{198}{231}$

21. What is the sum of $\frac{4}{15}$, $\frac{6}{7}$, and $\frac{3}{19}$? *Ans.* $1\frac{562}{1995}$.

22. What is the quotient of $\frac{37}{240}$ divided by $\frac{3}{8}$? *Ans.* $\frac{37}{90}$

23. Reduce the complex fraction $\frac{4\frac{1}{7}}{8}$ to its simplest form. *Ans.* $\frac{29}{56}$.

24. What is the value of $\frac{3}{17}$ of a mile? *Ans.* 1 fur. 16 rods, $7\frac{13}{17}$ feet.

25. Can the vulgar fraction $\frac{3}{11}$ be accurately expressed in decimals? *Ans.* It cannot.

26. How many places of decimals will be required to express $\frac{18}{32}$? *Ans.* 4 places.

27. Find the compound repetend equivalent to $\frac{1}{26}$. *Ans.* $0{\cdot}0\dot{3}8461\dot{5}$

28. Find the perfect repetend arising from $\frac{1}{47}$.

Ans. $\begin{cases} 0{\cdot}\dot{0}212765957446808510638\,2 \\ 978723404255319148936\dot{1}7. \end{cases}$

29. Convert 0·3756 into a vulgar fraction. *Ans.* $\frac{939}{2500}$.

30. Convert $\frac{371}{629}$ into a continued fraction.

$$Ans.\ \cfrac{1}{1+\cfrac{1}{1+\cfrac{1}{2+\cfrac{1}{3+\cfrac{1}{1+\cfrac{1}{1+\cfrac{1}{7+\cfrac{1}{2.}}}}}}}}$$

31. Find some of the approximative values of the continued fraction $\cfrac{1}{3+\cfrac{1}{3+\cfrac{1}{3+\cfrac{1}{3+, \&c.}}}}$

Ans. $\frac{0}{1}$, $\frac{1}{3}$, $\frac{3}{10}$, $\frac{10}{33}$, $\frac{33}{109}$, &c.

32. What must be the length of a thread which will wind spirally about a cylinder of 4 feet in circumference, and 60 feet in length, the distance between each turn of the thread being 1 foot? *Ans.* 247·38, &c. feet.

33. If A can perform a piece of work in 10 days, B in 12 days, C in 16 days, then how many days will be required for all together to perform the work? *Ans.* $4\frac{4}{59}$.

34. A shepherd, in the time of war, fell in with a party of soldiers, who plundered him of half his flock, and half a sheep over; afterwards a second party met him, who took off half of what he had left, and half a sheep over; and, soon after this, a third party met him, and used him in the same manner, and then he had only five sheep left. It is required to find what number of sheep he had at first. *Ans.* 47 sheep.

35. Four persons, A, B, C, and D, spent 20 shillings in company, when A proposed to pay $\frac{1}{3}$, B $\frac{1}{4}$, C $\frac{1}{5}$, and D $\frac{1}{6}$ part; but when the money came to be collected, they found it was not sufficient to answer the intended purpose. The question then is, to find how much each person must contribute to make up the whole reckoning, supposing their several shares to be to each other in the proportion above specified.

Ans. A must contribute $\frac{20}{57}$s.
B " " $\frac{15}{57}$s.
C " " $\frac{12}{57}$s.
D " " $\frac{10}{57}$s.

36. A stationer sold quills at 10 s. 6 d. a thousand, by which he cleared $\frac{1}{3}$ of this money; but, growing scarce, he raised them to 12 s. a thousand. What did he clear per cent. by the latter price? *Ans.* 71$\frac{3}{7}$ per cent.

37. How much can a person give for an annuity of $400, which has to run 12 years, if the interest be reckoned at 3 per cent.? *Ans.* $3981·602.

38. A debt, due at this present time, amounting to $1200, is to be discharged in seven yearly and equal payments. What is the amount of one of these payments, if the interest be calculated at 4 per cent.?

Ans. $199·931.

39. An usurer lent a person $600, and drew up for the amount a bond of $800, payable in 3 years, bearing no interest. What did he take per cent., if compound interest be taken into consideration?

Ans. 10·06424 per cent.

40. A debtor, being unable to pay his debts, amounting to $12950, at once, agrees with his creditors to discharge this sum by monthly instalments, viz.: $600 the 1st month,

and each succeeding month $50 more than the preceding one. In how many months will he have discharged his whole debts, and how much does he pay the last month?

Ans. In 14 months, and $1250.

41. A person dying, leaves half of his property to his wife, one sixth to each of two daughters, one twelfth to a servant, and the remaining $600 to the poor. What was the amount of his property? *Ans.* $7200.

42. We know, from natural philosophy, that any body which falls *in vacuo*, passes, in the first second, through a space of $16\frac{1}{12}$ feet; and, in each succeeding second, $32\frac{1}{6}$ feet more than in the one immediately preceding. Now, if a body has been falling 20 seconds, how many feet will it have fallen the last second, and how many in the whole time? *Ans.* $627\frac{1}{4}$ feet, and $6433\frac{1}{3}$ feet.

43. An estate of $7500 is to be divided between a widow, two sons, and three daughters, so that each son shall receive twice as much as each daughter, and the widow herself $500 more than all the children. What was her share, and what the share of each child?

Ans. { Widow's share, . . $4000.
Each son's, ... $1000.
Each daughter's, . $ 500. }

44. Three soldiers, in a battle, made $96 booty, which they wished to share equally. In order to do this, A, who made the most, gave B and C as much as they already had; in the same manner, B then divided with A and C, and, after this, C with A and B. If, then, by these means, the intended equal division is effected, how much booty did each soldier make?

Ans. A, $52; B, $28; and C, $16.

45. Two carpenters, twenty-four journeymen, and

eight apprentices, received, at the end of a certain time, \$144. The carpenters received \$1 per day, each journeyman half a dollar, and each apprentice 25 cents. How many days were they employed? *Ans.* 9 days.

46. A man, to please his children, brings home a number of apples, and divides them as follows: To the first and eldest of his children, he gives the half of the whole number, less 8; to the second, the half of the remainder, again diminished by 8, and he does the same with the third and fourth. After this, he gives the 20 remaining apples to the fifth. How many apples did he bring home? *Ans.* 80.

47. A farmer being asked how many sheep he had, answered that he had them in five fields; in the first he had $\frac{1}{4}$, in the second $\frac{1}{6}$, in the third $\frac{1}{8}$, in the fourth $\frac{1}{12}$, and in the fifth 450. How many had he? *Ans.* 1200.

48. I once had an untold sum of money lying before me. From this, I first took away the 3d part, and put in its stead \$50. A short time after, I took from the sum, thus augmented, the 4th part, and put again in its stead, \$70. I then counted my money, and found \$120. What was the original sum? *Ans.* \$25.

49. After paying away $\frac{1}{4}$ and $\frac{1}{5}$ of my money, I had \$66 left in my purse. How many dollars were in it at first? *Ans.* \$120.

50. A countryman brings his eggs to market, and first sells 4 more than the half of them; then he goes further, and sells half of the remainder, and 2 over. Now 6 eggs more than half of the remainder are stolen from him, and, dissatisfied about this loss, he returned to his village with the 2 eggs which remained in his basket How many eggs did he take to town? *Ans.* 80 eggs.

51. A person goes to a tavern with a certain sum of money in his pocket, where he spends 2 shillings; he then borrows as much money as he had left, and, going to another tavern, he there spends 2 shillings also; then borrowing again as much money as was left, he went to a third tavern, where, likewise, he spent 2 shillings, and borrowed as much as he had left; and, again spending 2 shillings at a fourth tavern, he then had nothing remaining. What had he at first? *Ans.* 3 s. 9 d.

52. A cistern can be filled by 3 pipes; by the first in $1\frac{1}{3}$ hours, by the second in $3\frac{1}{3}$ hours, and by the third in 5 hours. In what time will this cistern be filled when all 3 pipes are open at once? *Ans.* In 48 minutes.

53. Into what parts must 36 be divided, so that $\frac{1}{2}$ of the first, $\frac{1}{3}$ of the second, and $\frac{1}{4}$ of the third, may be all equal to each other? *Ans.* 8, 12, and 16.

54. A dog pursues a hare. Before the dog started, the hare had made 50 paces, and this is the distance between them at first. The hare takes 6 paces to the dog's 5, and 9 of the hare's paces are equal to 7 of the dog's. How many paces can the hare take before the dog overtakes her? *Ans.* 700 paces.

55. A person wishes to dispose of his horse by lottery. If he sells the tickets at $2 each, he will lose $30 on his horse; but if he sells them at $3 each, he will receive $30 more than his horse cost him. What was the number of tickets, and what did the horse cost him?

Ans. { The number of tickets was 60.
The horse cost him $150.

56. A person had two barrels, and a certain quantity of wine in each. In order to have an equal quantity in each, he poured out as much of the first cask into the

second as it already contained; then again he poured out as much of the second into the first as it then contained; and, lastly, he poured out again as much from the first into the second as there was still remaining in it. At last, he had 16 gallons of wine in each cask. How many gallons did they contain originally?

Ans. The first, 22; the second, 10 gallons.

57. What is the sum of the cubes of the seven following fractions: $\frac{420}{525}$, $\frac{497}{525}$, $\frac{504}{525}$, $\frac{510}{525}$, $\frac{511}{525}$, $\frac{513}{525}$, and $\frac{516}{525}$? *Ans.* 6.

58. A gentleman hires a servant, and promises him, for the first year, only $60 in wages; but, for each following year, $4 more than for the preceding. How much will the servant receive for the 17th year of his engagement, and how much for all 17 years together?

Ans. $124, and $1564.

59. A general, wishing to draw up his regiment into a square, tried it in two ways. The first time he had 39 men over; the second time, having extended the side of the square by one man, he wanted 50 men to complete the square. What was the number of soldiers in the regiment? *Ans.* 1975 soldiers.

60. From a sum of money, $50 more than the half of it is first taken away; from the remainder, $30 more than its fifth part; and again, from the second remainder, $20 more than its fourth part. At last there remained only $10. What was the original sum? *Ans.* $275.

61. Hiero, king of Sicily, ordered a crown to be made, containing 63 oz. of pure gold; but, suspecting that the workmen had debased it by using part silver, he recommended the detection of the fraud to the famous Archimedes, who, putting it into water, found that it

displaced 8·2245 cubic inches of water. He next found that a cubic inch of gold weighed 10·36 ounces, and a cubic inch of silver weighed 5·85 ounces; then, from this data, he calculated the proportions of gold and silver of which the crown was composed. What must have been his result?

Ans. { 2·5447 oz. of gold to every 2·1434 oz. of silver.

62. The number 48 is divided into two parts, such that if the greater part be multiplied by 24, and the less part by 12, the difference of their products will be 504. What are the parts? *Ans.* 30, and 18.

63. A man and his son together worked 50 days. The man received $1 per day, and the son $0·75. Their whole wages amounted to $44. How many days did each work?

Ans. { The man worked 26 days.
The son " 24 "

64. Three men, A, B, and C, together work 60 days, and all receive $48; A received $0·70 per day, B $0·80, and C $0·90. How many days did each work?

Ans. { A and C worked each the same number of days, and B worked the balance of the 60 days, be the same more or less.

65. Two men, A and B, are on a straight road, on the opposite sides of a gate, and distant from it 308 yards and 277 yards, respectively, and travel each towards the original station of the other. How long must they walk so that they shall be equally distant from the gate, if B goes 2 yards per second, and A $2\frac{1}{3}$ yards per second?

Ans. 1 minute, 33 seconds.

66. Every thing being supposed as in the preceding

question, at what time will each be at the same distance from the original station of the other as the other is from his? *Ans.* $4\frac{1}{2}$ minutes after starting.

67. Suppose the elastic power of a ball which falls from a height of 100 feet, to be such as to cause it to rise 50 feet, or one half the height from which it fell, and to continue in this way, diminishing the height to which it will rise, in geometrical progression, till it come to a state of rest, how far will it have moved?

Ans. 300 feet.

68. What is the square of 12890625?

Ans. 166168212890625.

NOTE.—In this question, it will be observed that the square of the above number ends with the same set of figures as the number itself; and this must hold good for any power of the above number. The same is true of the number 109376. Either of these numbers can be carried to any extent, by prefixing the proper figures. These are the only numbers which possess this remarkable property.

APPENDIX.

In this Appendix we shall, by using the Roman notation, number the different portions the same as the Articles in the body of the work, which they are designed to explain or enlarge upon.

II. The particular form of many of the symbols employed to indicate the relations of numbers, as well as the different operations which are required to be performed upon them, are no doubt entirely arbitrary. But some of them are easily traced to their true origin, and a reason is readily seen why they were employed. Under this article we propose to make a few remarks on what may be properly called the philosophy of arithmetical symbols. Not that we presume to know positively the reasons for the particular form of all symbols, but it is thought a few suggestions and queries on this subject might not be uninteresting to the inquiring mind.

The most natural method which could be devised for indicating that one quantity or number was greater than another is the symbol >, or <, so placed between the quantities or numbers compared as to open towards the greater, showing that in passing from the angle of this symbol to the opposite side, there is an increase. In this sense, this same symbol is used in music.

In conformity with this use of the above symbol we have $1>99 cents; 45 cents<$1; &c. The symbol $, employed to denote dollars is no doubt derived from a combination of the letters U and S, which, taken together, form the abbreviation for United States. If a compound letter be formed by uniting these letters, it will at once be seen how small a modification it is necessary to make in it so as to obtain the symbol $. And when we consider that this kind of currency, having a dollar for its unit, is peculiarly that of the United States, the explanation seems quite satisfactory.

Having agreed to employ the symbol $>$, or $<$, to denote inequality, we may ask how must it be modified when placed between two equal quantities so as to become the symbol of equality? In this case, there is no reason why this symbol should be placed with its openings towards the one quantity rather than the other, since neither is larger than the other. It ought to open equally towards each; hence it must of necessity become $=$, which is the symbol of equality.

The addition of one quantity to another might very concisely be represented by the union of two straight lines. But in what manner shall these two lines be united? When numbers are united by addition, the first may be said to be added to the second, or the second may be said to be added to the first, so that the two terms united have mutual relations with each other. So also when two numbers are united by multiplication, they are each factors, and have mutual relations. And moreover since the operations of multiplication are so nearly allied to those of addition, it would seem natural to represent multiplication also by the union of two lines so placed as to be mutually related. If two equal lines mutually bisect each other at right angles, it would form a symbol which might be appropriately used to denote either addition or multiplication. When these lines are so placed as that one may be horizontal and the other perpendicular, as +, it is the symbol of addition. When this symbol is so placed that the lines are oblique as ×, the symbol denotes multiplication.

Since the symbol +, denotes the union of two numbers by addition, if we take one away, or perform the operation of subtraction, this operation might be indicated by the symbol + after one of the lines, as for instance the vertical one, had been taken away, so as to give — for the symbol of subtraction.

The symbol ÷, for division, was probably suggested by the idea of separating or dividing by means of a line passing through the quantity, leaving a portion on each side of the straight line, which portions are represented by the two dots placed on the two sides of this straight line. If we substitute numbers for those dots, considering the one above the line to correspond with the dividend, and the one below the line to correspond with the divisor, we shall thus have, in the case of 7 divided by 5, the expression $\frac{7}{5}$, which is in strict accordance with the notation for a vulgar fraction, which shows that a vulgar fraction is but another method of denoting a division where the numerator corresponds with the dividend, and the denominator with the divisor.

If in the proportion 6 : 3 : : 8 : 4, we draw a straight horizontal line between the two dots separating the first and second terms, as well as between the two dots which separate the third and fourth terms ; also draw two horizontal lines joining the four dots between the second and third terms, we shall have

6÷3 :=: 8÷4,

which is read, 6 divided by 3 is equal to 8 divided by 4, and this is precisely what the first expression,

6 : 3 : : 8 : 4, meant.

The symbol $\surd$, which denotes a root, is without doubt derived from the letter *r*, which is the initial of *root.*

And may not the symbol ∴, which usually denotes *therefore,* have been formed from the consideration that in a syllogism there are three essential parts or propositions, which are indicated in this symbol by the three dots placed in the above symmetrical form?

V. *The difference between two numbers which are composed of the same digits, the digits being placed in any order whatever, is exactly divisible by* 9.

For two numbers, being composed each of the same digits, will, when divided by 9, each leave the same remainder, hence their difference must be an exact multiple of 9, that is, it must be exactly divisible by 9.

If from any number we subtract the sum of its digits, the remainder will be exactly divisible by 9.

For we have already shown that any number divided by 9, will give the same remainder as will be found by dividing the sum of its digits by 9, therefore the difference between any number and the sum of its digits is exactly divisible by 9.

From the above property, a very interesting arithmetical puzzle is deduced. Thus, you request a person to write down any number he may choose, then direct him to subtract from it the sum of all its digits; this done, request him to cancel or obliterate one of the digits of the result; after which, request him to give the sum of the remaining figures, and then you will be able to say just what figure was obliterated.

For, had no figure been obliterated, the sum of the digits must have been divisible by 9, therefore the figure which it is necessary to add to the result which he gives you, to make it a multiple of 9, is the figure that was obliterated. If a nine or zero had been obliterated, the result would still be a multiple of 9, and you could then only say that it was either 0, or 9, that had been cancelled, but you could not tell which.

The number 11 possesses some remarkable properties, which we will now proceed to point out.

$$\begin{aligned}
10^1 &= 10 = 11-1\\
10^2 &= 100 = 99+1\\
10^3 &= 1000 = 1001-1\\
10^4 &= 10000 = 9999+1\\
10^5 &= 100000 = 100001-1\\
&\text{\&c.} \quad \text{\&c.} \quad \text{\&c.}
\end{aligned}$$

From the above method of separating the different powers of 10, we see that if 1 is added to the odd powers of 10, the results will be divisible by 11; also, if 1 be subtracted from the even powers of 10, the results will also be divisible by 11, since the numbers 11, 99, 1001, 9999, 100001, &c., are each divisible by 11.

If we take any number, as 78534, it may, by the aid of what has just been shown, be decompounded as follows: $78534 = 70000 + 8000 + 500 + 30 + 4$.

But

$$\begin{aligned}
70000 &= 7\times 10000 = 7\times(9999+1) = 7\times 9999+7\\
8000 &= 8\times 1000 = 8\times(1001-1) = 8\times 1001-8\\
500 &= 5\times 100 = 5\times(99+1) = 5\times 99+5\\
30 &= 3\times 10 = 3\times(11-1) = 3\times 11-3\\
4 &= 4 = 4 = +4
\end{aligned}$$

Therefore, $78534=7\times 9999+8\times 1001+5\times 99+3\times 11$ $+(7-8+5-3+4)$.

Each of the expressions 7×9999, 8×1001, 5×99, 3×11, being exactly divisible by 11, it follows that the number 78534 is exactly divisible by 11 when the expression $7-8+5-3+4$ is. *That is, any number is divisible by* 11, *when the difference between the sum of the digits occupying the odd places, counting from the right-hand towards the left, and the sum of the digits occupying the even places, is divisible by* 11.

If then we take the difference between the sum of the digits occupying the odd places, and the sum of those occupying the even places, and divide it by 11, we shall obtain the same remainder as will be found by dividing the number itself by 11.

This method of determining whether a number is divisible by 11 may be simplified as follows:

Subtract the first figure from the second, the result from the third, this result from the fourth, and so on. The final result, or this result increased by 11, if it is negative, will be the remainder required.

Thus, what is the remainder when 1642915 is divided by 11? We commence at the left-hand and say, 1 from 6, 5; 5 from 4, -1; -1 from 2, 3; 3 from 9, 6; 6 from 1, -5; -5 from 5, 10; so that 10 is the remainder. Take the number 348531, we say 3 from 4, 1; 1 from 8, 7; 7 from 5, -2; -2 from 3, 5; 5 from 1, -4, to which add 11, we have 7 for the remainder.

By this method the pupil will see that subtracting a *minus*, or negative number, is the same as adding it, considered as positive, which is in accordance with the algebraic rule for the subtraction of negative numbers.

When a number is divided by 9, or when the sum of its digits is divided by 9, the remainder is called its excess of nines.

Hence, in addition, the sum of the excess of nines of the different numbers added, is equal to the excess of nines of the sum.

In subtraction, if we subtract the excess of nines of the subtrahend from the excess of nines of the minuend, the remainder will be the excess of nines of the difference. If the excess of the minuend is less than that of the subtrahend, we must first increase it by 9 before subtracting.

In multiplication, the product of the excesses of the factors is equal to the excess of the total product.

In division, the excess of the divisor multiplied by the excess of the quotient and added to the excess of the remainder is equal to the excess of the dividend.

By the help of the foregoing relations, we are enabled to verify the operations of the four fundamental rules of arithmetic.

Similar methods of verification could be employed in reference to the number 11.

The operation of multiplication may be readily verified by using any number whatever, in accordance with the following property: If the product of any two factors be divided by any number, the remainder will be the same as would be found by first finding the remainders in reference to the same divisor of the respective factors, taking the product of these remainders, and then finding the remainder of this product. As an example, what is the remainder by 7 of the product of 365 by 89? The remainder of 365 by 7 is 1; the remainder of 89 by 7 is 5; therefore the remainder of the product is $1\times 5=5$. Again, what is the remainder by 7 of 377 multiplied by 87? The remainder of 377 by 7 is 6; the remainder of 87 by 7 is 3. Therefore the remainder sought is the same as the remainder of $6\times 3=18$ taken in reference to 7, which is the remainder of 4.

VI. Much time has been spent, and much ingenuity displayed by many distinguished mathematicians in attempting to find a sure and simple means of discovering whether a number is a prime. Among other methods we will mention the following:

Take the continued product of all the natural numbers until we reach the number which is a unit less than the number under consideration, to this product add 1, if the result is divisible by the number, then will that number be a prime.

Thus, take the number 7, we have the product $1\times 2\times 3\times 4\times 5\times 6=720$, to which add 1, we have 721, which is exactly divisible by 7, consequently, by the above, the number 7 is a prime.

Although this method is strictly accurate in theory, it is of no practical benefit, on account of the great labor of performing the multiplication of many factors, when the number is of considerable magnitude.

By this principle we may readily solve the following problem:

A gentleman asked a shepherd what number of sheep

he had, who answered, that when he numbered them 2 and 2, 3 and 3, 4 and 4, 5 and 5, 6 and 6, there remained one in each case; but when he numbered them 7 and 7, there remained 0; what number of sheep had he?

The number $2\times3\times4\times5\times6=720$, will divide exactly by 2, 3, 4, 5, and 6, since it is composed of these factors, hence, adding 1, the result, 721, will, when divided by either of the numbers 2, 3, 4, 5, or 6, give one for a remainder; but it is exactly divisible by 7, since 7 is a prime.

The least multiple of 2, 3, 4, 5, and 6, is 60. Now find the least multiple of 60, which with 1 added will become divisible by 7, which multiple is 300, to which adding 1, we have 301, which also satisfies the above question. In a similar manner other answers might be found.

The following methods of deducing primes by a series of numbers, according to a fixed law, deserves notice:

Write the series of even numbers, 2, 4, 6, 8, 10, 12, 14, 16, 18, 20, 22, &c., then commencing with 17, add it to 2, the first term of this series, and we obtain 19, which added to 4, the second term of the series, gives 23, which in turn added to 6, the third term, gives 29; thus continuing we find the new series

17, 19, 23, 29, 37, 47, 59, &c.

Each term of this series is a prime, and this will be found to hold good until we reach the 16th term, which will be the composite number 17×17. The 17th term will give the composite number 17×19. After passing the 17th term, the 18th, 19th, 20th, &c., yield primes.

If to our series of even numbers we had commenced

with 41 instead of 17, we should have found the following primes :

41, 43, 47, 53, 61, 71, &c.,

which fails at the 40th and 41st terms, which are respectively the composite numbers 41×41, 41×43.

If to the arithmetical progression 2, 6, 10, 14, 18, 22, &c., we commence with 29, and add as before, we shall find the series of primes,

29, 31, 37, 47, 61, 79, 101, &c.,

which holds good until we reach the 28th term, which is the composite number 59×29, after which it again gives primes.

XVII. When a number terminates on the right with a zero, it is a whole number of tens, and is therefore divisible by 2 as well as by 5.

When a number terminates with 5, it must be divsible by 5, since the tens are divisible by 5, and the units being 5, are also divisible by 5.

When the number expressed by the two right-hand figures of any number is divisible by 4, the whole number is divisible by 4. For the number is composed of a whole number of hundreds and the number expressed by the two right-hand figures. The hundreds are always divisible by 4, and, by supposition, the part expressed by the two right-hand figures is also divisible by 4, consequently the whole number is divisible by 4.

When the number expressed by the three right-hand figures of any number is divisible by 8, the whole number will be divisible by 8. For the number is composed of a whole number of thousands together with the num ber expressed by the three right-hand figures. The thousands are always divisible by 8, and, by supposition, the number expressed by the three right-hand figures is also divisible by 8, consequently the whole number is divisible by 8.

From the above, we are able to determine whether any given number is divisible by either of the numbers 2, 3, 4, 5, 6, 8, 9, 10, 11, 12.

Thus if we wish to know whether a number is divisible by 2, we must see whether it ends on the right with a zero or with the figure 2, in either case it is divisible by 2.

If we wish to know whether it is divisible by 3, we must see if the sum of its digits is divisible by 3, or see whether there is no excess of threes, if so, then the number is divisible by 3.

If we wish to know whether it is divisible by 4, we see whether the number expressed by its two right-hand figures is divisible by 4, if so, the number is divisible by 4.

If we wish to know whether it is divisible by 5, we see whether it ends on the right with a zero or a 5, if so, it is divisible by 5.

If we wish to know whether a number is divisible by 6, we first see if it is divisible by 2, we also see if it is divisible by 3, if so, then it is divisible by 6, since 6 equals 2 times 3.

If we wish to know whether it is divisible by 8, we see if the number expressed by its three right-hand figures is divisible by 8, if so, the number is divisible by 8.

If we wish to know whether a number is divisible by 9, we see whether the sum of its digits is divisible by 9, if so, the number is divisible by 9.

If we wish to know whether a number is divisible by 10, we notice whether the number end on the right with a zero, if so, the number is divisible by 10.

If we wish to know whether a number is divisible by 11, we see if the difference of the sum of the digits of the even places, and those of the odd places, is zero, or a multiple of 11, which may be most readily done by subtracting the first or left-hand digit from the second, and this result from the third, and the result now obtained from the fourth, and so on, as explained under V. of this Appendix. If then the difference is either zero, or a multiple of 11, the number is divisible by 11.

If we wish to know whether a number is divisible by 12, we see whether it is divisible by 3, we also see if it is divisible by 4, if so, it will be divisible by 3 times 4, or by 12.

There is no simple practical method of determining whether a number is divisible by 7. The best way, in this case, is to make the trial by the method of short division.

By the application of the above rules, which in practice will be found to be very simple, we may at once abbreviate or reduce vulgar fractions, when any of these factors are common to the numerator and denominator.

If the numerator and denominator are both primes, the fraction is in its lowest terms. Also, if either the numerator or denominator is a prime, and not a divisor of the other, the fraction is in its lowest terms.

When a decimal fraction is expressed by writing its

denominator, it becomes a vulgar fraction whose denominator is some power of 10, and therefore contains no prime factors except 2 and 5, hence if the numerator is not divisible by 2 or 5, the fraction is in its lowest terms.

LIII. We have seen that *Complementary repetends*, which include the *Perfect repetends*, arise from expressing the reciprocals of primes by the aid of decimals. All prime numbers will not give rise to complementary repetends, they will all give *Simple repetends*, except 2 and 5. The following table, shows the number of places of decimals in each period of the simple repetend, arising from the primes up to 1051, except the primes 2 and 5, whose reciprocals are accurately expressed by decimals. Those in the table marked with C give complementary repetends, and those marked with P are not only complementary repetends, but are also perfect repetends. It will be seen, by inspecting this table, that when the number of decimal places is one less than the prime, it is not only complementary but also perfect.

At first view the pupil might imagine the labor of forming this table to be exceedingly great, on account of the great number of places in some of the periods; this labor would truly be immense were we obliged to find these decimals by the usual method, but if we employ the process explained under Art. **45**, the work is rapidly performed. Indeed the number of places in a period may be found even without actually finding the decimal figures, by a method which is rather simple, but which would require considerable space to explain.

Primes.	No. of dec. places.	Primes.	No. of dec. places.	Primes.	No. of dec. places.	Primes.	No. of dec. places.	Primes.	No. of dec. places.
3	1	163	81	P 367	366	P 593	592	P 823	822
P 7	6	P 167	166	C 373	186	599	299	827	413
C 11	2	173	43	P 379	378	C 601	300	C 829	276
C 13	6	P 179	178	P 383	382	C 607	202	839	419
P 17	16	P 181	180	P 389	388	613	51	853	213
P 19	18	191	95	397	99	C 617	88	P 857	856
P 23	22	P 193	192	C 401	200	P 619	618	C 859	26
P 29	28	C 197	98	C 409	204	631	315	P 863	862
31	15	199	99	P 419	418	C 641	32	C 877	438
37	3	C 211	30	C 421	140	643	107	C 881	440
41	5	P 223	222	431	215	P 647	646	883	441
43	21	227	113	P 433	432	C 653	326	P 887	886
P 47	46	P 229	228	439	219	P 659	658	907	151
53	13	P 233	232	443	221	C 661	220	C 911	450
P 59	58	239	7	C 449	32	C 673	224	919	459
P 61	60	C 241	30	C 457	152	C 677	338	C 929	464
67	33	C 251	50	P 461	460	683	341	P 937	936
71	35	P 257	256	C 463	154	C 691	230	P 941	940
C 73	8	P 263	262	467	233	P 701	700	947	473
79	13	P 269	268	479	239	P 709	708	P 953	952
83	41	271	5	P 487	486	719	359	C 967	322
C 89	44	277	69	P 491	490	P 727	726	P 971	970
P 97	96	C 281	28	P 499	498	733	61	P 977	976
C 101	4	283	141	P 503	502	C 739	246	P 983	982
C 103	34	C 293	146	P 509	508	P 743	742	991	495
107	53	307	153	C 521	52	751	125	C 997	166
P 109	108	311	155	523	261	757	27	1009	252
P 113	112	P 313	312	P 541	540	C 761	380	1013	253
C 127	42	317	79	547	91	C 769	192	P 1019	1018
P 131	130	C 331	110	C 557	278	773	193	P 1021	1020
C 137	8	P 337	336	563	281	787	393	1031	103
C 139	46	347	173	C 569	284	797	199	P 1033	1032
P 149	148	C 349	116	P 571	570	C 809	202	1039	519
151	75	C 353	32	P 577	576	P 811	810	C 1049	524
C 157	78	359	179	587	293	P 821	820	P 1051	1050

We have shown that the reciprocal of all prime numbers, except 2 and 5, when expressed by decimals, give repetends whose number of places of figures cannot exceed the units, less one, contained in the prime number. When the number of places in the period is less than the units of the prime, after one has been subtracted, it must be a multiple of this number. So that if the process of decimating the reciprocal of any prime number, except 2 and 5, be carried to as many decimals, less one, as there are units in the prime, the remainder must of necessity be a unit. That is, if we divide the power of 10, which is denoted by an exponent which is one less than our prime divisor, the remainder will be 1. Hence, if from this power we subtract 1, the remainder will be exactly divisible by this prime. When 10 is raised to any power, it will consist of 1 placed to the left of as many zeros as there are units in the exponent, and if we subtract 1, the result will be a number denoted by as many nines as there was zeros. *Hence, any prime divisor, except* 2 *and* 5, *is exactly contained in a number which consists of as many successive nines as there are units, less one, in the prime.*

As examples, 3 is a divisor of 99; 7 is a divisor of 999999; 11 is a divisor of 9999999999; 13 is a divisor of 999999999999; and so on.

As each of these expressions is divisible by 9, which contains only the prime 3, their quotient, which will consist of a succession of 1s, must be divisible by the same primes as before, excepting in the case of the prime 3, which is a factor of 9. *That is, any prime divisor, except* 2, 3, *and* 5, *is exactly contained in a number which consists of as many successive* 1s *as there are units less one in the prime.*

As examples, 7 is a divisor of 111111; 11 is a divisor of 1111111111; 13 is a divisor of 111111111111; and so on.

If these numbers, which consist of a succession of 1s, are multiplied by any one of the nine digits, the products will consist of numbers composed of the same digits taken as many times in succession as there were 1s in the number multiplied. This multiplication cannot impair their divisibility. *Hence, in general, any prime divisor, except* 2, 3, *and* 5, *is exactly contained in a number which consists of as many successive digits of any kind as the units in the prime divisor.*

As examples, 7 is a divisor of the numbers 111111, 222222, 333333, 444444, 555555, 666666, 777777, 888888, and 999999. 11 is a divisor of the numbers 1111111111, 2222222222, 3333333333, &c.

We will in this place mention the following additional properties of numbers, the reasons for which may be readily discovered by the student:

The sum of any number of *even* numbers, is an even number.

The difference of any two *even* numbers, is an even number.

The sum of an even number of *odd* numbers, is an even number.

The sum of an odd number of *odd* numbers, is an odd number.

The difference of two *odd* numbers, is an even number.

The sum, or difference, of an *even* number and an *odd* number, is odd.

The product of an *even* and *odd* number, is even.

If an *even* number be divisible by an *odd* number, the quotient will be even.

The product of any number of factors is *even*, if any one of the factors is even.

Any number which is a measure of two numbers is also a measure of their sum, their difference, and their product.

LXIII. Since we consider 30 days to the month, and 12 months to the year, it might, at first view, seem that we made the year to consist of $30\times 12=360$ days; so that the interest obtained by our rule would be to the true interest, in the ratio of 365 to 360; but a more careful investigation will show that no such error is committed; that in fact, in most cases, a very slight difference, if any, can exist. For when we wish the interest on months, whether we consider a month of 30 days, or any other number of days, all will be right, so long as a month is regarded as $\frac{1}{12}$ of a year, for then evidently the interest for one month is one-twelfth of the interest for one year. So that the only difference which can possibly occur is in reference to the interest on whatever odd days there may be.

This rule, while it is very near the truth, is an exceedingly simple rule, and therefore a most excellent one in practice.

Since the interest of a certain sum at seven per cent. for one year is $\frac{7}{100}$ of the principal, it will in one day be

$\frac{1}{365}$ of $\frac{7}{100}$ of the principal, or by cancelling the 7 of the numerator against a part of the 365 of the denominator, we find $\frac{1}{52\frac{1}{7}}$ of $\frac{1}{100}$ of the principal. If for $\frac{1}{52\frac{1}{7}}$, we use its close approximate value $\frac{1}{52}$, we shall see that the interest on a given sum for one day may be found by dividing the given principal by 52 and removing the decimal point two places towards the left, for removing the decimal point two places to the left is equivalent to dividing by 100. Hence, the interest on a given sum at 7 per cent. for any number of days may be found by the following concise

RULE.

Multiply the given principal by the number of days, divide the product by 52, and remove the decimal point two places towards the left.

As an example, find the interest of $59·23 for 13 days, at 7 per cent.

Operation.

```
     59·23
        13
    ------
    17769
    5923
    ------
52)769·99(14·80,
    52
    ---
    249
    208
    ---
     419
     416
     ---
       39
```

Removing the decimal point two places towards the left, we have $0·148 for the interest.

LXVII. If A owes me $100, due at the end of 3 months, and $100, due at the end of 9 months, and I accept his note for $200, at 6 months, do I gain or lose by this operation?

Since I have delayed the payment of the first $100 for 3 months beyond the time it was due, I of course lose the interest of $100 for 3 months; but on the other hand I receive the second $100 three months before it was due, by which I gain the discount of $100 for 3 months. But the discount of a certain sum of money at a certain rate per cent. is always less than the interest of the same sum at the same rate per cent., and for the same time. Therefore I lose by the change.

And in general, if a person accept, for the sum of several debts, due at different periods, an obligation at an equated time, as found by the usual rules, he loses. This loss is usually so small, and the usual rule for finding the equated time so simple, as to cause it to be preferred. Still we have thought it in place here to point out its defects. The new rule deduced by considering the present values of the several debts is strictly accurate.

In connection with this subject I will add the following questions, which may be found in the Elementary Arithmetic:

1. If I buy a certain number of oranges at 3 cents each, and as many more at 5 cents each, and sell them at 4 cents each; do I gain or do I lose?

Ans. I neither gain nor lose.

2. If I buy a certain number of oranges at 3 for one cent, and as many more at 5 for one cent, and sell them at 4 for one cent; do I gain, or do I lose?

Ans. I lose.

3. If I expend a certain sum of money for oranges at 3 cents each, and as much more money for other oranges at 5 cents each, and sell them at 4 cents each; do I gain, or do I lose?

Ans. I gain.

A careful consideration of these three questions, which at first might appear nearly equivalent, will aid very much in the right understanding of some of the conditions of successive operations of arithmetic, which we shall presently explain.

In the first example, the 4 cents for which each orange was sold is an arithmetical mean between 3 and 5, and therefore exceeds the 3 cents, the price of the first lot of oranges, just as much as it falls short of 5 cents, the price of the second lot; and as there was just as many oranges at 3 cents each as there were at 5 cents each, there can be neither gain nor loss.

In the second example, the $\frac{1}{4}$ of a cent for which each orange was sold is not an arithmetical mean between $\frac{1}{3}$ and $\frac{1}{5}$, the respective prices paid for the different lots of oranges; but $\frac{1}{4}$ is the reciprocal of the mean of 3 and 5. The first lot of oranges having cost $\frac{1}{3}$ of a cent each, by selling them at $\frac{1}{4}$ of a cent each, we lose on each orange $\frac{1}{3}-\frac{1}{4}=\frac{1}{12}$ of a cent. On the second lot there was a gain of $\frac{1}{4}-\frac{1}{5}=\frac{1}{20}$ of a cent on each orange; hence the loss is greater than the gain.

In the third example, the number of oranges bought in the first lot is equal to one-third the number of cents expended, and in the second lot the number was one-fifth of the money expended; so that more than one-half of the whole number of oranges were bought for 3 cents each, while less than one-half cost 5 cents each; consequently, as they were all sold at 4 cents each, there must have been a gain.

From these remarks, in connection with the foregoing examples, we see that, *the reciprocal of the mean of two numbers, as of the numbers* 3 *and* 5, *is not the same as the mean of the reciprocals of the same numbers.*

The first, that is, the reciprocal of the mean of 3 and 5, is $\dfrac{1}{\frac{1}{2}(3+5)}=\frac{1}{4}$.

The second, that is, the mean of the reciprocals of 3 and 5 is $\frac{1}{2}\left(\frac{1}{3}+\frac{1}{5}\right)=\frac{4}{15}$.

Here, then, we discover that it makes a difference whether we first take the mean, and then take the reciprocal of that mean, or whether we first take the reciprocals of the numbers, and then take the mean of those reciprocals.

Similar results arise from other operations; as *the square root of the sum of two quantities is different from the sum of the square roots of the same quantities.*

Thus, the square root of the sum of 9 and 16 is

$$\sqrt{9+16}=\sqrt{25}=5.$$

The sum of the square roots of 9 and 16 is

$$\sqrt{9}+\sqrt{16}=3+4=7.$$

On the other hand there are many arithmetical operations which yield the same result, in whatever order of succession they are performed, as for instance, the roots of powers, or the powers of roots. Thus, the cube root of the second power of 8 is $(8^2)^{\frac{1}{3}}=(64)^{\frac{1}{3}}=4$. The second power of the cube root of 8 is $(8^{\frac{1}{3}})^2=(2)^2=4$.

When several successive operations, which are required to be performed upon a number or numbers, are of such a nature as to give the same result, whatever may be their order of succession, they may with propriety be called *homogeneous operations.*

But, when these operations are such as to give different results, when the order of succession is different, they may be called *heterogeneous operations.*

Taking of roots, raising to powers, multiplying, dividing, and reciprocating, are homogeneous operations.

So with addition and subtraction.

But addition or subtraction, taken in connection with either of the preceding mentioned operations, become heterogeneous operations.

As an example of homogeneous operations, we will give the following, which gives six varieties, all leading to the same result:

1. The reciprocal of the third root of the second power of 8 is $\frac{1}{(8^2)^{\frac{1}{3}}}=\frac{1}{4}$.

2. The reciprocal of the second power of the third root of 8, is $\frac{1}{(8^{\frac{1}{3}})^2}=\frac{1}{4}$.

3. The second power of the third root of the reciprocal of 8, is $[(\frac{1}{8})^{\frac{1}{3}}]^2=\frac{1}{4}$.

4. The second power of the reciprocal of the third root of 8, is $(\frac{1}{8^{\frac{1}{3}}})^2=\frac{1}{4}$.

5. The third root of the second power of the reciprocal of 8, is $[(\frac{1}{8})^2]^{\frac{1}{3}}=\frac{1}{4}$.

6. The third root of the reciprocal of the second power of 8, is $(\frac{1}{8^2})^{\frac{1}{3}}=\frac{1}{4}$.

There are many algebraic operations, which by continually repeating them give results which recur in periods. In arithmetic, we have seen an instance of this in the case of decimating certain vulgar fractions, which give a decimal value consisting of a period of figures which continually recur to infinity. The process of reciprocating a number several times in succession, leads to results which recur in periods.

Thus, the reciprocal of 4 is $\frac{1}{4}$; the reciprocal of the reciprocal of 4, or the reciprocal of $\frac{1}{4}$ is 4; the reciprocal of the reciprocal of the reciprocal of 4 is $\frac{1}{4}$; and, the reciprocal of the reciprocal of the reciprocal of the reciprocal of 4 is 4; and so on, where the results are 4, $\frac{1}{4}$, 4, $\frac{1}{4}$, 4, &c., constantly recurring in periods.

Taking the reciprocal of a number is the same as dividing a unit by that number, and again taking the reciprocal of this result, is the same as dividing a unit by this result, or which is the same, it consists in dividing a unit by the reciprocal of this number, but dividing by a number is the same as multiplying by its reciprocal, therefore, taking the reciprocal of the reciprocal of a number is the same as multiplying a unit by the number, which evidently gives for a result the number itself, which explains why these successive results recur in periods.

The following is a very curious result of the principle of successive operations: *Any root of any number being continually taken, the successive results continually approximate to a unit.*

For, if the number is greater than a unit, its root will be smaller than the number, and therefore it will be nearer a unit than the quantity; and for the same reason the root of this root will be still nearer the value of a unit;

and thus continuing to take the root of the last result, each result must be nearer 1 than the preceding result.

If the original number is less than a unit, its root will exceed the number, and therefore be nearer the value of a unit than the number; and for a similar reason the root of this root will be still nearer 1. Hence, we may conclude that any root of any number being successively taken an infinite number of times, the result must become 1. After reaching 1, which it could not do short of an infinite number of extractions, it could not possibly depart from this unit value, since all roots as well as all powers of 1 are 1.

LXXI. Perhaps there is nothing which requires more to be well understood by all business men, than the principles and rules in relation to interest of money.

According to all the methods now in use in different countries, as well as the various rules adopted by our different states, there is wide room for different individuals to arrive at very different results in computing the interest on many bills of obligation. That which gives rise to the most difficulty, is the distinction between *simple* and *compound* interest. By the laws of this country, a loan for any period, even though it exceeds one year, is not allowed any more than simple interest; while at the same time the loan may be made for as short a period as we please, and at the end of said period the interest may be added to the principal and the result again loaned as a new principal; by which means it may be made to draw even a greater interest than the ordinary compound interest. So long as the statute of the land has laws to prevent usury, it ought to provide some certain and infallible method for computing interest.

Another common source of perplexity in computing interest, is in the case of bills of obligation, where payments are made at various periods, some being made at intervals less than one year, and others at longer intervals. So great has this source of perplexity been, that our different states have adopted their distinct rules in regard to the method of computing interest in such cases.

It is obvious that so long as the unit of time is a *finite* quantity, as one year, no rule can be devised which shall in all cases be equitable to both parties, for when the payments are made before the end of the year, they must affect the parties different from what they would when made after they became due. By the many ways in which contracts are drawn, new cases are almost daily presented, which require much labor as well as skill to determine the exact rate per cent. per annum received.

I see no way by which such difficulties can be avoided, so long as a finite portion of time is allowed before the interest is considered as due. It may be asked, why is a man's interest any more due at the end of the year, than the half of it at the end of six months? or than any fractional part of a year's interest which may have accrued at any other period? The true way, in my opinion, is to consider the interest due at any and all periods of time; or in other words, that a sum of money at interest should be constantly augmenting; that is, I would have all sums of money, whether for great or small portions of time, at compound interest, the interest being added in at the end of every *instant*. This would give a unity to all cases of casting interest; no misunderstanding could then possibly arise as to the amount of interest due. I shall show that it is not difficult to compute interest tables on this principle. The per cent. per annum may be so taken as to make the interest due at the end of one year, the same as in the case of simple interest.

I do not suppose that such a method of casting interest will ever come into general use; all I wish is, to show that such a method is practicable, and if adopted would at once settle all disputes in respect to *usury*. Interest would then be a uniformly increasing quantity, not limited to any particular epoch for receiving its increments.

If we consider instantaneous compound interest to be taken at the rate per cent. per annum, as expressed by the second column of the following table, it will produce the same result at the end of the year as would be found by using the corresponding per cent. given by the first column, when simple interest is considered.

Rate per cent, Simple interest.	Rate per cent, instantaneous Compound interest.	Difference between these rates per cent.
0·03	0·0295587	0·0004413
0·035	0·0344013	0·0005987
0·04	0·0392206	0.0007794
0·045	0·0440169	0·0009831
0·05	0·0487902	0·0012098
0·055	0.0535408	0·0014592
0·06	0.0582690	0·0017310
0·065	0.0629748	0·0020252
0·07	0.0676587	0·0023413

The following table makes a comparison between the simple interest at 7 per cent., and the compound instantaneous interest at 6·76587 per cent. per annum, for portions of time not exceeding one year.

Days.	Amount of $1 at simple interest, the rate per cent. per annum being 7.	Amount of $1 at compound interest, the interest being compounded at the end of every instant,—the rate per cent, per annum being 6·76587.	Difference between the simple interest and the compound interest.
1	1·000192	1·000183	0·000009
2	1·000384	1·000371	0·000013
3	1·000575	1·000556	0·000019
4	1·000767	1·000742	0·000025
5	1·000959	1·000927	0·000032
6	1·001151	1·001113	0·000038
7	1·001342	1·001298	0·000044
8	1·001534	1·001484	0·000050
9	1·001726	1·001670	0·000056
10	1·001918	1·001855	0·000063
20	1·003836	1·003714	0·000122
30	1·005753	1·005576	0·000177
40	1·007671	1·007442	0·000229
50	1·009589	1·009311	0·000278
60	1·011507	1·011884	0·000323
70	1·013425	1·013060	0·000365
80	1·015342	1·014940	0·000402
90	1·017260	1·016823	0·000437
100	1·019178	1·018709	0·000469
200	1·038356	1·037769	0·000587
300	1·057534	1·057185	0·000349
365	1·070000	1·070000	0·000000

By the foregoing table we see that the greatest difference, as given in the fourth column, is $0·000587, which corresponds to 200 days, or a little more than half a year; the greatest difference, when more accurately found is $0·000592, which corresponds with 0·5025 of a year. This difference is less than $\frac{6}{10}$ of a mill on one dollar. If this instantaneous compound interest is deemed too great, the rate per cent. per annum may be taken as much smaller as we please. All that we wish to show is, that this method of considering instantaneous compound interest is practicable, and may be safely and readily employed. It is true, the tables to be employed in this case require for their construction a higher knowledge than can be given by arithmetic alone. The same is true of many of our tables in frequent use, such as logarithmic tables, astronomic tables, &c.

Before closing this article, we will show that the compound interest, calculated strictly by the usual rule, becomes less than the simple interest, when the time is less than one year.

The amounts, at compound interest, of $1 for 1, 2, 3, 4, &c., years, at 7 per cent., are represented by $(1·07)^1$, $(1·07)^2$, $(1·07)^3$, $(1·07)^4$, &c., respectively. From which we see that this amount is a power of 1·07, having for its exponent a number equal to the number of years. Now when the time becomes less than one year, as for instance, ½ of a year, the expression for the amount of $1 will become $(1·07)^{\frac{1}{2}}$, the value of which is the square root of 1·07, which root we find to be 1·0344, nearly. So that, by this process, we find the compound interest of $1 for 6 months, at 7 per cent., to be $0·0344, while the simple interest on the same sum for the same time, at the same rate per cent., is $0·035, which shows that in this case the compound interest for a period less than one year, is less than the simple interest. And by a similar process it may be shown to be the same for other rates per cent., and other periods of time, which are less than one year. When the compound interest is to be computed for a given number of years and a fraction of a year, the usual method is to compute the amount at compound interest for the whole number of years, and then on this amount to compute the simple interest for the fraction of the year, which gives a trifle more than would be found by the method of logarithms, where the mixed number denoting the years and the fraction of a year is taken for the exponent. This method of computing compound interest by logarithms is explained in Treatise on Algebra.

LXXIV. The square of the sum of any number of numbers may be found as follows: By rule under Art. **4**, we know that the square of the sum of any two numbers, as 6+8, is equal to $6^2+2\times6.8+8^2$, which result may be thus expressed:

The square of the sum of two numbers is equal to the square of the first number, plus twice the product of the first number into the second, plus the square of the second.

If we wish the square of the sum of three numbers, as 6+8+9, we may unite the first and second by means of a parenthesis, thus, for 6+8+9, we may make use of (6+8)+9, and now regarding 6+8 as one number, the preceding rule for the sum of two numbers, will apply to (6+8)+9, that is, the square of 6+8+9 is equal to the square of (6+8), plus twice the product of (6+8) into 9, plus the square of 9. But the square of 6+8 has already been shown to be, the square of 6, plus twice the product of 6 into 8, plus the square of 8. Hence, the square of 6+8+9 is equal to the square of 6, plus twice the product of 6 into 8, plus the square of 8, plus twice the product of the sum of 6 and 8 into 9, plus the square of 9. Or in general terms,

The square of the sum of three numbers is equal to the square of the first number, plus twice the product of the first number into the second, plus the square of the second; plus twice the product of the sum of the first two into the third, plus the square of the third.

Continuing in this way, we could show that, *the square of the sum of any number of numbers is the square of the first number, plus twice the product of the first number into the second, plus the square of the second; plus twice the product of the sum of the first two into the third, plus the square of the third; plus twice the product of the first three into the fourth, plus the square of the fourth; plus twice the product of the first four into the fifth, plus the square of the fifth; and so on.*

We will now apply this general rule to a few examples.

1. $(2+3)^2=2^2+2\times 2.3+3^2$.
2. $(5+7)^2=5^2+2\times 5.7+7^2$.
3. $(3+4+5)^2=3^2+2\times 3.4+4^2+2\times(3+4).5+5^2$.
4. $(5+6+7)^2=5^2+2\times 5.6+6^2+2\times(5+6).7+7^2$.
5. $(7+8+9)^2=7^2+2\times 7.8+8^2+2\times(7+8).9+9^2$.
6. $(35)^2=(30+5)^2=30^2+2\times 30.5+5^2$.
7. $(47)^2=(40+7)^2=40^2+2\times 40.7+7^2$.
8. $(365)^2=(300+60+5)^2=300^2+2\times 300.60+60^2+2\times(300+60).5+5^2$.
9. $(487)^2=(400+80+7)^2=400^2+2\times 400.80+80^2+2\times(400+80).7+7^2$.

The above method of squaring a number consisting of the sum of two or more numbers, is elegantly illustrated geometrically as follows:

The square ABCD may be enlarged to the square AEKF, by the addition of the two equal rectangles BG, and DH, whose lengths are each equal to the side AB of the original square, and whose widths are equal to BE, the quantity by which the side of the square has been augmented, also a little square, CGKH, whose side is the same as the width of one of the equal rectangles.

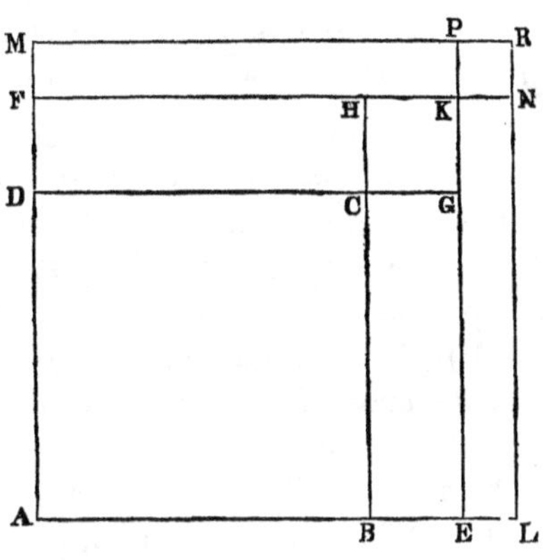

Again, the square AEKF, having its side increased by EL, becomes augmented by the two rectangles EN, FP, and the little square KR. Thus we might continue to augment the square last found by the addition of two equal rectangles, and a little square ; the length of each rectangle being equal to the side of the square which is to be augmented, and the width equal to the quantity by which the side of the square is increased, also, the side of the little square will be the same as the width of one of the rectangles.

We will now endeavor to discover a general rule for the cube of the sum of any number of numbers. If we wish the cube of $6+8$, we will first take its square, which is $6^2+2\times 6{\cdot}8+8^2$, multiplying this by $6+8$, by the rule under Art. 4, we obtain $6^3+3\times 6^2.8+3\times 6.8^2+8^3$,

That is, *the cube of the sum of two numbers is, the cube of the first number, plus three times the product of the square of the first number into the second, plus three times the product of the first into the square of the second, plus the cube of the second.*

If we wish the cube of the sum of three numbers, as $6+8+9$, we may unite the first and second by means of a parenthesis, thus, for $6+8+9$, we may make use of $(6+8)+9$, and regarding $(6+8)$ as one number, we find that the cube of $(6+8)+9$ or the cube of $6+8+9$ is equal to the cube of $(6+8)$ plus three times the product of the square of $(6+8)$ into 9, plus three times the product of $(6+8)$ into the square of 9, plus the cube of 9. But the cube of $6+8$, has already been shown to be equal to the cube of 6, plus three times the product of the square of 6 into 8, plus three times the product of 6 into the square of 8, plus the cube of 8. Hence the cube of $6+8+9$ is equal to the cube of 6, plus three times the square of 6 into 8, plus three times 6 into the square of 8, plus the cube of 8; plus three times the square of the sum of 6 and 8 into 9, plus three times the sum of 6 and 8 into the square of 9, plus the cube of 9. *And in general, we have the cube of the sum of any number of numbers equal to the cube of the first number, plus three times the square of the first number into the second, plus three times the first into the square of the second, plus the cube of the second; plus three times the square of the sum of the first two into the third, plus three times the sum of the first two into the square of the third, plus the cube of the third; plus three times the product of the square of the sum of the first three into the fourth, plus three times the sum of the first three into the square of the fourth, plus the cube of the fourth, and so on.*

Thus,

$$(2+3)^3=2^3+3\times 2^2.3+3\times 2.3^2+3^3.$$

$$(5+7)^3=5^3+3\times 5^2.7+3\times 5.7^2+7^3.$$

$$(5+6+7)^3=5^3+3\times 5^2.6+3\times 5.6^2+6^3$$
$$+3\times(5+6)^2.7+3\times(5+6).7^2+7^3.$$

$$(365)^3=(300+60+5)^3=300^3+3\times 300^2.60$$
$$+3\times 300.60^2+60^3+3\times(300+60)^2.5$$
$$+3\times(300+60).5^2+5^3.$$

This method of giving the cube of the sum of two or more numbers may be illustrated geometrically as follows:

Let the cube, which is represented by the figure 1, be augmented by three equal rectangular parallelopipedons, which we will call *slabs*, and it will appear as represented by figure 2. Let this second solid be further augmented by three other equal rectangular parallelopipedons which we will call *corner pieces*, and it will take the form, as represented by figure 3. Finally increase this last solid by the *little cube* at the corner, and we shall obtain the perfect cube represented by figure 4.

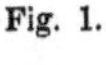

Fig. 1.

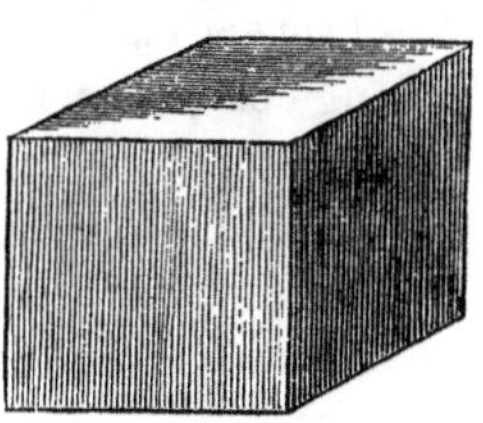

Fig. 2.

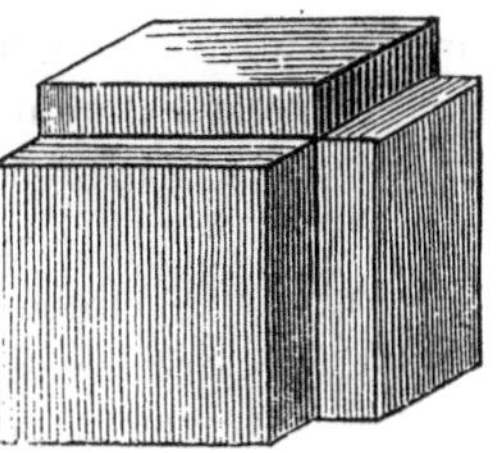

The length and breadth of one of these equal *slabs* is the same as the side of the original cube, the thickness is the quantity by which the side of the cube is augmented.

Fig. 3.

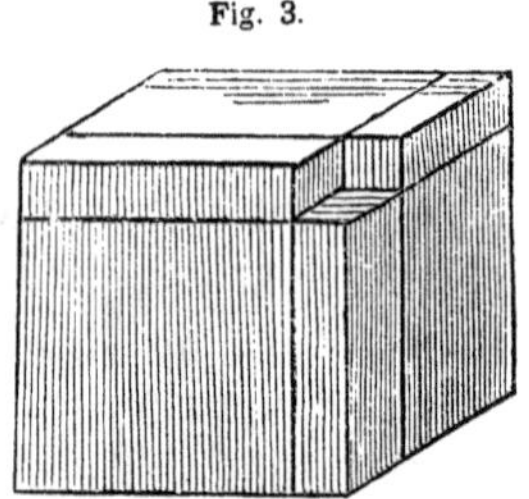

The length of one of the equal *corner pieces* is the same as the side of the original cube, its thickness and width is the same as the thickness of the *slabs.*

The side of the little cube at the corner is the same as the thickness of the *slabs* or *corner pieces.*

Fig. 4.

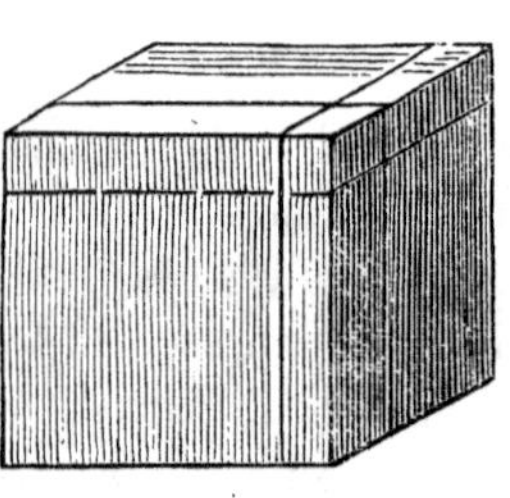

In the same way the cube denoted by figure 4 may be augmented by three equal *slabs*, three equal *corner pieces*, a *little corner cube*, and when thus augmented it will still be a perfect cube.

We will now endeavor to explain more fully the reason of the rules for extracting the square root, and cube root of numbers.

Let it be required to extract the square root of 531441. For the sake of simplicity we will suppose we are required to find the number of feet in the side of a square, so that its area shall contain 531441 square feet.

We know that the number denoting the square root of 531441, or the number of feet in the side of our sought square must consist of three places of figures, since the number 531441 is greater than the square of 100 and less than the square of 1000. The greatest square which does not exceed 53 is 49, hence the figure in the hundredth's place is 7, that is, the side of the square sought, is greater than 700 and less than 800 feet. If we suppose a square whose side is 700 feet, it will contain 490000 square feet, which subtracted from 531441 square feet leaves 41441 square feet, which is called the first dividend. It is necessary to increase the side of our supposed square, so that its area shall be augmented by 41441 square feet. A square may be increased by equal additions on all its four sides and still retain its square form, or it may be increased equally on any two adjacent sides. The latter method being the simplest, is the plan we shall follow.

If, then, we suppose two equal rectangles added upon any two adjacent sides of the square of 700 feet, there will still be required a small square whose side is equal to the width of these additional rectangles, in order to form a perfect square. Now as the area of these rectangles is by far the largest portion of the additional part, and since the length of each rectangle is 700 feet, both together will be 1400 feet, the trial divisor, it follows that if we divide 41441 by 1400, we shall obtain the approximate width of these rectangles. Performing the division, we find a quotient between 20 and 30, hence, the figure in the ten's place is 2.

If to the sum of the lengths of the two rectangles, which is 1400, we add the side of the additional square, which is 20, being the same as width of rectangles, we shall have 1420, the true divisor, which, multiplied by 20, gives 28400 for the number of square feet thus added. Subtracting 28400 from 41441, we find 13041 for the second dividend. So that our second square, whose side is 720 feet, is less than the sought square by 13041 square feet, hence it is necessary, again, to increase this last square, whose side is 720 feet, by two equal rectangles, the length of each being 720 feet, and of a certain width, and a small square whose side is equal to the width of these additional rectangles. The sum of the lengths of these two rectangles is 1440, the second trial divisor. Therefore, dividing 13041 by 1440 we find a quotient between 9 and 10, therefore, the units which express the width of the additional rectangles is 9 feet, consequently the side of the little additional square is also 9 feet, which added to 1440, the sum of the length of the rectangles, gives 1449, the second true divisor, this, multiplied by 9, gives 13041 square feet for the whole addition, which being subtracted from 13041 leaves no remainder, so that a square whose side is 729 feet, contains just 531441 square feet, and is therefore the side of the square sought.

Had the number, whose root it was required to find, been such as to lead to more than three places of figures in the root, the process would not differ from the above, only in being more lengthy. Each addition being found by the consideration of augmenting the square last formed, by two equal rectangles, whose lengths are each equal to the side of the square, and an additional square whose side is equal to the width of those rectangles. So that each additional figure of the root is found with the same certainty as the successive figures in the quotient of an operation in division.

If the above process be carefully compared with the work of Example 1, under Case I. of Extraction of the Square Root, it is believed the pupil cannot fail of obtaining a clear and satisfactory reason of the rule there given.

Let us endeavor to extract the cube root of 387420489, which is Example 1, under Case I., Extraction of the

Cube Root, to which work we shall, in the following, make reference. For the sake of simplicity we will suppose we are required to find the number of feet in the side of a cube, so that its volume shall be 387420489 cubic feet.

We know that the number denoting the cube root of 387420489, or the number of feet in the side of our sought cube, must consist of three places of figures, since the number 387420489 is greater than the cube of 100, and less than the cube of 1000. The greatest cube which does not exceed the left-hand period, 387, is 343, hence, the figure in the hundredth's place is 7, that is, the side of the cube sought is greater than 700, and less than 800 feet. If we take the cube whose side is 700 feet, the area of one of its faces will be $(700)^2 = 490000$ square feet, and its volume will be $490000 \times 700 =$ 343000000 cubic feet, which taken from 387420489 leaves 44420489 cubic feet. Hence, it is necessary to increase the side of this cube, so that its volume may be augmented by 44420489 cubic feet, which is called the first dividend. Now we know that a cube may be increased by three equal rectangular parallelopipedons, which we have called *slabs*, (see page 315, of this Appendix,) three other equal rectangular parallelopipedons called *corner pieces*, and a *little cube;* and when thus increased it will still be a perfect cube. Since the volume of the three *slabs* is by far the largest portion of the additions, and since the surface of each *slab* is the same as the face of the cube, viz. $(700)^2 = 490000$ square feet, the three together will be $3 \times (700)^2 = 1470000$ square feet, which has been called the first trial divisor, it therefore follows, that if we divide 44420489 by 1470000, we shall obtain the approximate thickness of these *slabs*. Performing this division, we find a quotient between 30 and 40, but as the divisor so used is always less than the surface of all the additional parts, by the surface of the three *corner pieces* and *little cube*, our quotient may sometimes be too large, it can never be too small; in this case, 30 is too large, therefore we must consider the true quotient as comprised between 20 and 30, so that the figure in the ten's place is 2.

Hence, the thickness of the *slabs* is 20 feet, this is also the width of the *corner pieces*, their lengths being the same as the side of the cube, 700, so that the surface of the three corner pieces is $3 \times 700 \times 20 = 42000$. Also the side of the *little cube* is 20, one of its faces is $(20)^2 = 400$, hence, the entire surface of all the additions is $1470000 + 42000 + 400 = 1512400$ equal the true divisor, which being multiplied by 20 gives 30248000 cubic feet, which subtracted from 44420489, leaves 14172489 cubic feet, which still remain to be added to the cube whose side is 720.

Using, as before, three times the square of 720, which is $3 \times (720)^2 = 1555200$, which is the surface of the three new *slabs*, making a second *trial divisor*, which we find is contained in 14172489 between 9 and 10 times, therefore the unit figure of our root is 9, which is the thickness of the new *slabs*, and the width of the new *corner pieces*, also the side of the new *little cube*, hence, the entire surface of the additions is $3 \times (720)^2 + 3 \times (720) \times 9 + 9^2 = 1574721$ equal *true divisor*, which multiplied by 9 gives 14172489 for the whole number of cubic feet added this second time, which being subtracted from the second dividend, 14172489, leaves no remainder, so that a cube whose side is 729 feet, contains just 387420489 cubic feet, and is therefore the side of the cube sought.

Had the number, whose root it was required to find, been such as to give more than three places of figures in the root, the process would not have differed from the above, only being more lengthy, each additional figure being found by means of a trial divisor which is three times the surface of a face of the cube already obtained. Each successive operation, when a new figure is found, augments the cube last obtained by three *slabs*, three *corner pieces*, and a *little cube*. In the arrangement of the work under Example 1, which has already been referred to, we have omitted the zeros on the right, for the sake of simplicity; for the same reason, the work is arranged in two distinct columns, by which means the operation of multiplying, &c., is made quite simple. By means of these auxilliary columns the work bears a close analogy to the new method of solving numerical cubic equations, as given by Mr. Horner, of *Bath, England*. The use of auxiliary columns becomes very apparent in the extraction of roots of the higher orders, as the fifth root, the seventh root, &c. The above rule for the cube root may be beautifully illustrated by means of the ordinary blocks prepared for the common rule.

In squaring a number, we multiply the right-hand digit into itself, thus obtaining the right-hand digit of the product, or of the square; hence, the right-hand digit of a perfect square must arise from squaring some one of the digits. The right-hand digit of the squares of the nine digits will end with one of the following digits, 1, 4, 5, 6, or 9. Therefore every square number must end with one of the digits 1, 4, 5, 6, 9, or else with an even number of zeros. Hence, any number terminating on the right with either of the digits 2, 3, 7, 8, or with an odd number of zeros, cannot have its square root accurately found.

The right-hand digit of the cubes of the nine digits will end with all the different digits, so that we cannot, by the terminal figure of a number, pronounce that its cube root cannot be found, except when the number ends with a number of zeros which is not a multiple of 3, in which single case we know that the cube root cannot be accurately found.

When a number is divided by 5, we shall obtain, for a remainder, one of the following numbers, 1, 2, 3, 4, unless the number is a multiple of 5, in which case it is exactly divisible by 5.

If a number is divisible by 5, its square will obviously be divisible by 5.

If a number give 1 for a remainder, when divided by 5, it may be considered as composed of a multiple of 5, plus 1; consequently when squared by the method explained under this article, page 311, its first and second terms will be multiples of 5, and the third term will be the square of 1, so that when a number is composed of a multiple of 5 plus 1, its square will give the same remainder, when divided by 5, as will be found by dividing the square of 1 by 5, which remainder is therefore 1.

And for a similar reason, when a number is composed of a multiple of 5, plus 2, the remainder found by dividing its square by 5, is the same as the remainder found by dividing the square of 2 by 5, which remainder is 4.

If a number is composed of a multiple of 5, plus 3, the remainder found by dividing its square by 5, will be the same as the remainder found by dividing the square of 3 by 5, which remainder is 4.

If a number is composed of a multiple of 5, plus 4, the remainder found by dividing its square by 5, will be the same as the remainder found by dividing the square of 4 by 5, which remainder is 1.

From which we discover, that any square number being divided by 5, will give either 1 or 4 for a remainder, unless it is a multiple of 5.

Hence, every square number is either a multiple of 5, *or else a multiple of* 5 *increased or decreased by* 1.

Any number divided by 7, will give one of the following remainders, 1, 2, 3, 4, 5, 6, unless it is a multiple of 7, in which case it is obviously divisible by 7.

When a number is not a multiple of 7, it may be regarded as composed of two parts, the first of which is a multiple of 7, plus one of the remainders, 1, 2, 3, 4, 5, 6.

Now, when a number of this kind is cubed by the method explained on page 315 of this Appendix, its cube will consist of four parts, namely, the cube of the first, plus three times the square of the first into the second, plus three times the first into the square of the second, plus the cube of the second; the first three of these parts contain a multiple of 7 as a factor, hence, the remainder will be found by dividing by 7 the last part, which is the cube of one of the numbers 1, 2, 3, 4, 5, 6. The cubes of 1, 2, 3, 4, 5, and 6, are respectively 1, 8, 27, 64, 125, and 216, dividing these cubes by 7, we find either 1 or 6 for the remainder.

From which we see that any cube number being divided by 7, will give either 1 or 6 for a remainder, unless it is a multiple of 7.

Therefore, every cube number is either a multiple of 7, *or else a multiple of* 7 *increased or diminished by* 1.

A TABLE

OF SQUARES, CUBES, SQUARE ROOTS, CUBE ROOTS, AND RECIPROCALS OF PRIMES, AS FAR AS 113.

No.	Square.	Cube.	Square Root.	Cube Root.	Reciprocal.
1	1	1	1·0000000	1·000000	1
2	4	8	1·4142136	1·259921	0·5
3	9	27	1·7320508	1·442250	0·3333333
5	25	125	2·2360680	1·709976	0·2
7	49	343	2·6457513	1·912931	0·1428571
11	121	1331	3·3166248	2·223980	0·0909091
13	169	2197	3·6055513	2·351335	0·0769230
17	289	4913	4·1231056	2·571282	0·0588235
19	361	6859	4·3588989	2·668402	0·0526316
23	529	12167	4·7958315	2·843867	0·0434783
29	841	24389	5·3851648	3·072317	0·0344828
31	961	29791	5·5677644	3·141381	0·0322581
37	1369	50653	6·0827625	3·332222	0·0270270
41	1681	68921	6·4031242	3·448217	0·0243902
43	1849	79507	6·5574385	3·503398	0·0232558
47	2209	103823	6·8556546	3·608826	0·0212766
53	2809	148877	7·2801099	3·756286	0·0188679
59	3481	205379	7·6811457	3·892996	0·0169492
61	3721	226981	7·8102497	3·936497	0·0163934
67	4489	300763	8·1853528	4·061548	0·0149254
71	5041	357911	8·4261498	4·140818	0·0140845
73	5329	389017	8·5440037	4·179339	0·0136986
79	6241	493039	8·8881944	4·290840	0·0126582
83	6889	571787	9·1104336	4·362071	0·0120482
89	7921	704969	9·4339811	4·464745	0·0112360
97	9409	912673	9·8488578	4·594701	0·0103093
101	10201	1030301	10·0498756	4·657009	0·0099010
103	10609	1092727	10·1488916	4·687548	0·0097087
107	11449	1225043	10·3440804	4·747459	0·0093458
109	11881	1295029	10·4403065	4·776856	0·0091743
113	12769	1442897	10·6301458	4·834588	0·0088496

By the aid of the foregoing table we may by multiplication determine the square root, cube root, and reciprocal of any number which does not contain a prime factor greater than 113.

Suppose we wish the square root of 365, we decompose 365 into the prime factors 5 and 73. Now, it is obvious that the root of the product of any number of factors is equal to the product of their roots, hence, the square root of 365 is equal to the square root of 5×73, which is equal to the square root of 5 multiplied by the square root of 73. By our table we find

$$
\begin{array}{r}
\sqrt{73} = 8{\cdot}5440037 \\
\sqrt{5} \ = 2{\cdot}2360680 \\
\hline
17{\cdot}0880074 \\
17088007 \\
2563201 \\
512640 \\
5126 \\
683 \\
\hline
\end{array}
$$

The multiplication is by the abridged method, as explained under Art. **39.**

$\sqrt{365} = 19{\cdot}1049731$ nearly.

For the cube root of the same number, we have the following work:

$$
\begin{array}{r}
\sqrt[3]{73} \ = 4{\cdot}179339 \\
\sqrt[3]{5} \ \ = 1{\cdot}709976 \\
\hline
4{\cdot}179339 \\
2925537 \\
37614 \\
3761 \\
292 \\
25 \\
\hline
\end{array}
$$

$\sqrt[3]{365} = 7{\cdot}146568$ nearly.

For the reciprocal of 365, we have as follows:

reciprocal of 73=0·0136986
" 5=0·2

0·00273972=reciprocal of 365.

This last work for finding the reciprocal of a composite number, reposes upon the principle, that the reciprocal of the product of any number of factors is equal to the product of their reciprocals.

The roots of composites are very readily obtained when the number can be separated into two factors, one of which can have its root accurately found: thus, the square root of $5537=49\times113$, is equal to 7 times the square root of $113=10·6301458\times7=74·4110206$ nearly.

The cube root of 664= the cube root of 8 times 83= 2 times the cube root 83= 2 times 4·362071=9·724142 nearly.

LXXXIII. This rule for finding the sum of all the terms of an infinite decreasing geometrical progression, has a beautiful application in the following problem, which is sometimes so stated as to appear paradoxical.

At what time next after 1 o'clock will the hour-hand of a clock overtake the minute-hand?

The distance which the hour-hand is forward of the minute-hand, is at first $\frac{1}{12}$ of the circumference of the dial plate. When the minute-hand shall have moved over this $\frac{1}{12}$, the hour-hand will have advanced $\frac{1}{12}$ of it, or $\frac{1}{12}$ of $\frac{1}{12}$ of the circumference, which is the distance which now separates them. Again, while the minute-hand moves over this last distance of $\frac{1}{12}$ of $\frac{1}{12}$ of the circumference, the hour-hand moves over $\frac{1}{12}$ of it, that is, $\frac{1}{12}$ of $\frac{1}{12}$ of $\frac{1}{12}=\frac{1}{12^3}$, which is the distance which now separates them. Again while the minute-hand moves over this last distance, the hour-hand advances $\frac{1}{12}$ of it, or over a distance $=\frac{1}{12}$ of $\frac{1}{12}$ of $\frac{1}{12}$ of $\frac{1}{12}$ of the circumference, which is the distance which now separates them. Hence, as the hour-hand always moves $\frac{1}{12}$ as far as the distance last moved by the minute-hand, it follows that the minute-hand cannot overtake the hour-hand. But we know that it does overtake it, inasmuch as it completes the circuit of the dial plate, and therefore must of necessity pass it. How then is this seeming contradiction to be explained?

Although we have assumed an infinite number of partial movements made by the hands, each succeeding move being $\frac{1}{12}$ of the preceding one, still the time required to make this *infinite* number of moves is *finite*.

These successive moves made by the minute-hand, considering the circumference of the dial plate as the unit, form this progression,

$$\frac{1}{12}, \frac{1}{12^2}, \frac{1}{12^3}, \frac{1}{12^4}, \frac{1}{12^5}, \&c.$$

This being a decreasing geometrical progression, whose ratio as well as first term is $\frac{1}{12}$, its sum is found to be $\frac{1}{11}$. Now, as the minute-hand can move through the whole circumference of the dial plate in one hour, it will require $\frac{1}{11}$ of one hour, which equals 5 minutes and $\frac{5}{11}$ of a minute to move over $\frac{1}{11}$ of the circumference of the dial plate, which is the exact time required for the minute-hand to overtake the hour-hand.

From which it appears that the foregoing fallacy consists in inferring that, because there was an infinite number of successive operations, it must require an infinite length of time to perform them, which, as we have seen, is not a legitimate conclusion.

Question 67, page 282, affords a second example of an infinite number of successive operations being performed in a finite time. For the number of bounds made by this elastic ball, provided it move in accordance with the law mentioned in the question, is *infinite,* since every time it falls it must bound, and every time it bounds it must of necessity fall.

We have already given 300 feet as the whole distance through which the ball moved. We will now show that the time required is finite.

The successive distances, expressed in feet, through which the ball fell, form the series 100, 50, 25, $12\frac{1}{2}$, &c., to an infinite number of terms. By the law of falling bodies, the distances fallen through are to each other as the squares of the times employed in falling; and $16\frac{1}{12}$ feet is the distance which a body falls in the first second of its descent; consequently, the times, expressed in seconds, which were required to fall through the respective distances which are given in the above series, form the following series: $\left(\frac{100}{16\frac{1}{12}}\right)^{\frac{1}{2}}$, $\left(\frac{50}{16\frac{1}{12}}\right)^{\frac{1}{2}}$, $\left(\frac{25}{16\frac{1}{12}}\right)^{\frac{1}{2}}$, &c., to an infinite number of terms. This is a geometrical series whose first term is $\left(\frac{100}{16\frac{1}{12}}\right)^{\frac{1}{2}}=10\left(\frac{12}{193}\right)^{\frac{1}{2}}$, and the ratio is $\sqrt{\frac{1}{2}}=\frac{1}{2}\sqrt{2}$. Therefore its sum, as found by rule under Article **83,** is $\frac{10\sqrt{\frac{12}{193}}}{1-\frac{1}{2}\sqrt{2}}$.

Again, the distances passed through in bounding form

the series 50, 25, 12½, &c., and the time required to make these bounds will be the sum of the geometricai progression $\left(\frac{50}{16\frac{1}{12}}\right)^{\frac{1}{2}}, \left(\frac{25}{16\frac{1}{12}}\right)^{\frac{1}{2}}, \left(\frac{12\frac{1}{2}}{16\frac{1}{12}}\right)^{\frac{1}{2}}$, &c. which sum is $\frac{\sqrt{50} \times \sqrt{\frac{12}{193}}}{1 - \frac{1}{2}\sqrt{2}}$. If this is added to the sum of the former series, we obtain

$$(10+\sqrt{50}) \times \frac{\sqrt{\frac{12}{193}}}{1-\frac{1}{2}\sqrt{2}} = (20+2\sqrt{50}) \times \frac{\sqrt{12}}{2\sqrt{193}-\sqrt{386}}$$

The value of this expression exceeds a little 14½, so that the whole time before the body comes to a state of rest is about 14½ seconds, or less than ¼ of a minute.

A question is often proposed in Mathematical geography, which leads to conclusions similar to the above. The question is this: Suppose the earth to be a perfect sphere, entirely covered with water, and that a vessel starting from the equator, sails without deviating from its course or velocity, always in the direction of North-east, when will it reach the north pole of the earth?

The answer is generally given, that it will never reach the pole. This is wrong, for although the track of the vessel is in the form somewhat of a spiral, which makes an infinite number of turns about the earth before reaching the pole, still the whole time required to perform these turns, or to pass over its whole track, is finite. The technical name of this curve is the *Loxodromic curve.*

POSITION.

There is a method by which many questions may be accurately wrought, and by which all others may be solved approximately. It consists in making one or more assumptions for the answer, and then from the error or errors thus arising, to deduce the true answer, or its approximate value, when only an approximation can be obtained.

This method is called *Position*, and sometimes it is called *the Rule of False*, or *the Rule of False Position*, or, which is far better, *the Rule of Trial and Error*. This rule admits of two varieties, *Single Position*, and *Double Position*.

In Single Position, only one assumption is required, while in Double Position two assumptions are necessary.

Single Position may be employed in the solution of questions, in which the required number is in any manner increased or decreased by any given part of itself; that is, when it is multiplied or divided by any given number.

Double Position must be used when the result obtained by increasing or decreasing the required number in any given ratio, is also increased or decreased by some number independent of the required number. Or when any root or power of the required number is either directly or indirectly given in the conditions of the question.

SINGLE POSITION.

From the above definition of Single Position, it follows that if the number assumed is only one-half the true number, the result will be only one-half of the result of the question. If the assumed number is twice as great as the true number, the result will be twice as great as the result of the question. And in general, the result obtained will be to the result of the question, as the assumed number is to the true number. Hence, we deduce for Single Position this

RULE.

Assume any convenient number, and perform on it the operations required by the question, then, as the result thus obtained, is to the result of the question, so is the assumed number to the true number required.

EXAMPLES.

1. Find a number such that being increased by one-half, one-third, one-fourth, and one-fifth of itself, the sum shall be 1644.

If we assume 120 for the number, its half will be 60, its third 40, its fourth 30, and its fifth 24. Hence, 120 increased by its half, its third, its fourth, and its fifth, becomes 274 for our result, and

274 : 1644 : : 120 : 720=the number sought.

2. A father bequeaths to his three sons $10700, in such a manner that the share of the first being multiplied by 5, that of the second by 6, and that of the third by 7, the products will be equal. What was each one's share?

Since the first multiplied by 5, the second by 6, and the third by 7, give equal products, it follows that their portions are as the reciprocals of these numbers, that is, as $\frac{1}{5}$, $\frac{1}{6}$, $\frac{1}{7}$. These fractions reduced to a common denominator give the following numerators: 42, 35, 30. We will therefore assume that the first had $42, the second

$35 and the third $30. Taking the sum, we find $107 for our result. Hence,

107 : 10700 : : $42 : $4200=portion of first son.
107 : 10700 : : $35 : $3500= do. second "
107 : 10700 : : $30 : $3000= do. third "

3. A, B, and C, joined their stock and gained $360, of which A received a certain sum, B received $3\frac{1}{2}$ times as much as A, and C as much as A and B together. What share of the gain had each?

Suppose A received $10, then as B received $3\frac{1}{2}$ times as much, he must have had $35, and as C had as much as A and B together, his portion must have been $45. Hence, all together received $90, which is our result.

Hence, 90 : 360 : : $10 : $40= A's gain.

Consequently, B's gain was $140, and C's gain $180, so that all together gained $360.

We will not extend the method of Single Position any further, since all questions under this rule may be solved by the succeeding rule for Double Position, or indeed without Position, by the method of Analysis.

DOUBLE POSITION.

The method of Double Position requires two assumed numbers, and depends upon the principle that the differences between the true and assumed numbers, are to each other as the differences between the result given in the question and the results arising from the assumed numbers. This principle is rigidly correct for such questions as, when solved by Algebra, would give rise to simple equations; but it is only approximately correct for all other questions. Admitting the above principle, we readily deduce the following

RULE.

Assume two different numbers, and perform on them separately the operations indicated in the question. Then, as the difference of the results thus obtained, is to the difference of the assumed numbers, so is the difference between the true result and either of the others, to the correction to be applied to the assumed number which gave this result. Add the correction to this number, if the corresponding result was too small; if too large, it must be subtracted.

When the question is of such a nature as to admit only of an approximate solution, we may for a second approximation assume the number already found for the first, and that one of the two first assumptions which was nearer the true answer, or any other number that may appear to be still nearer. In this way, by repeating the operation as often as may be necessary, the true result may be approximated to any assigned degree of accuracy.

The above method of approximation by Double Position is of very little value in the usual questions of Arithmetic, but it becomes of the greatest utility in Algebra, affording, in many cases, a very concise and convenient mode of approximating the roots of equations, and finding the values of unknown quantities in very complex expressions, without making the usual reductions.

EXAMPLES.

1. Required a number, from which if 2 be subtracted,

one-third of the remainder will be 5 less than half the required number.

First, assume the number to be 14, from which take 2, and one-third of the remainder is 4. This being subtracted from one-half of 14, the remainder is 3, the *first result.* Again, suppose the number 20, from which subtracting 2, one-third of the remainder is 6, which being taken from the one-half of 20, the remainder is 4, the *second result.* The difference of the results being 1, and the difference of the assumed numbers 6: and the difference between 5, the true result, and 4, the result nearest it, being 1, we have

$$1 : 6 :: 1 : 6.$$

And as the result 4 is too small, this correction of 6 must be added to 20, its corresponding assumed number, so that we have 26 for the number sought.

2. A and B together, agree to dig 100 rods of ditch for $100. That part of the ditch on which A was employed was more difficult of excavation than the part on which B was employed. It was therefore agreed that A should receive for each rod 25 cents more than B received for each rod which he dug. How many rods must each dig, and at what prices, so that each may receive just $50?

Since the average of the prices was $1, and there was $0·25 cents difference, we will, as a first assumption, take $1·13 for the price of a rod of A's portion, and consequently, the price of a rod of B's will be $0·88. As each received $50, we find that the number of rods dug by A was 50÷1·13=44·25 rods, nearly. The number of rods dug by B was 50÷0·88=56·82 rods, nearly, and they both together dug 101·07 rods, while the true result is 100.

As a second supposition, we will take \$1·14 for the price of a rod of A's portion, consequently \$0·89 will be the price per rod of B's portion; now, as before, we find that A dug $50 \div 1·14 = 43·86$ rods nearly. The number of rods dug by B was $50 \div 0·89 = 56·18$ rods, nearly; so that both together dug 100·04 rods, which result is very nearly correct. The difference of these results is 1·03, the difference of the assumed numbers is $1·14 - 1·13 = 0·01$; also the difference between the true result 100, and the second result, 100·04, is 0·04; therefore we have

$$1·03 : 0·01 :: 0·04 : 0·0004.$$

This correction 0·0004, being added to 1·14, which was the second assumed price per rod of the portion which A dug, gives \$1·1404, for a very close approximate value of the price per rod of A's portion. A still closer approximation might be found by repeating the process.

This last example may be found in the "Elements of Algebra," under Quadratic Equations. It is a good example for showing the value of the rule of Double Position as a method of approximation.

As the price of the whole 100 rods was just \$100, the average was \$1 per rod, and as the difference of the prices which A and B received per rod was $\frac{1}{4}$ of a dollar, it might at first be supposed that A's price was $1 + \frac{1}{8} = \frac{9}{8}$ dollars, and B's $1 - \frac{1}{8} = \frac{7}{8}$ dollars, but a still further investigation will make it evident that this question is similar to the second orange question, on page 301 of this Appendix; since 1 dollar is not the mean of the reciprocals of the prices $\frac{9}{8}$ and $\frac{7}{8}$. This mean is $\frac{64}{63}$, which being greater than 1, it follows that the price of a rod of A's portion must be greater than $\frac{9}{8}$ of a dollar, and B's as much greater than $\frac{7}{8}$ of a dollar.

We have been pretty full and explicit in the explanation of this question, since it is the type of a class of questions which have been looked upon by many as involving an absurdity.

ARITHMETICAL CURIOSITIES.

What is the least number of weights, and what are they, so that by the aid of the common balance we may weigh any number of integral pounds from 1 to 40?

By means of two weights we can, besides weighing the same number of pounds which they represent, weigh a number denoted by their sum and their difference.

Hence, one pound may be weighed by using a one-pound weight, or by using two weights whose difference is one pound. The best economy is no doubt to use a one-pound weight. Now, the largest weight which can be employed in connection with this one-pound weight in order to weigh 2 pounds, is a three-pound weight. The greatest number of pounds which can be weighed with these two weights is 4 pounds. And in order to weigh 5 pounds it will be necessary to introduce a nine-pound weight, then by these three weights we can weigh 13 pounds, and in order to weigh 14 pounds we must have a twenty-seven-pound weight. We have thus made use of four different weights, namely, a one-pound weight, a three-pound weight, a nine-pound weight, and a twenty-seven-pound weight. By combining the weights as follows we shall be able to weigh any number of integral pounds from 1 to 40:

1; $3-1=2$; 3; $3+1=4$; $9-3-1=5$; $9-3=6$; $9+1-3=7$; $9-1=8$; 9; $9+1=10$; $9+3-1=11$; $9+3=12$; $9+3+1=13$; $27-9-3-1=14$; $27-9-3=15$; $27+1-9-3=16$; $27-9-1=17$; $27-9=18$; $27+1-9=19$; $27+3-9-1=20$; $27+3-9=21$; $27+3+1-9=22$; $27-3-1=23$; $27-3=24$; $27+1-3=25$; $27-1=26$; 27; $27+1=28$; $27+3-1=29$; $27+3=30$; $27+3+1=31$; $27+9-3-1=32$; $27+9-3=33$; $27+9+1-3=34$; $27+9-1=35$; $27+9=36$; $27+9+1=37$; $27+9+3-1=38$; $27+9+3=39$; $27+9+3+1=40$.

A sea-captain, on a voyage, had a crew of 30 men, half of whom were blacks. Being becalmed on the passage for a long time, their provisions began to fail, and the captain became satisfied that, unless the number of men was greatly diminished, all would perish of hunger before they could reach any friendly port. He therefore proposed to the sailors that they should stand in a row on deck, and that every ninth man should be thrown overboard, until one-half of the crew were thus destroyed. To this they all agreed. How should they stand so as to save the whites?

Ans. { WWWWBBBBBWWBWWWBWBBWW BBBWBBWWB.

This result might have been easily ascertained beforehand by the captain, by the following method: Place 30 letters, as the letter W, it being the initial of white, in a row; then commence counting off every ninth one, which may be canceled, when 15 have in this way been

29

canceled, their places may be supplied with the letter B, the initial of black.

A and B have an eight-gallon cask full of wine, and wish to make an equal division of it, but have only two empty vessels with which to do it, one of which contains 5 gallons, and the other 3. By what means shall they accomplish this division?

	8 gall. cask.	5 gall. cask.	3 gall. cask.
1st	8	0	0
2d	3	5	0
3d	3	2	3
4th	6	2	0
5th	6	0	2
6th	1	5	2
7th	1	4	3

Explanation. In the first place there is 8 gallons in the 8 gallon cask, and the others are empty. 2d, the 5 is filled from the 8, leaving 3 gallons in the 8. 3d, the 3 is filled from the 5, leaving 2 gallons in the 5. 4th, the 3 gallons of the 3 is emptied into the 8, which already contains 3 gallons, thus giving 6 gallons in the 8, and 2 in the 5. 5th, the 2 gallons in the 5 are poured into the 3d. 6th, the 5 is filled from the 8, leaving 1 in the 8. 7th, the 3, which contains 2 gallons, is now filled from the 5, thus leaving 4 gallons in the 5.

The following, which is a similar question to the last, we will leave to the ingenuity of the pupil:

Three persons have stole a vessel of balsam containing 24 ounces, and have three vessels containing 5, 11, and 13 ounces respectively; in what manner must they proceed to effect the distribution, so that each may receive an equal portion?

A very interesting arithmetical puzzle, is to desire a person to think of any three numbers, each of which is less than 10; then direct him to multiply the first by 2 and to add 5 to the product; to multiply this sum by 5 and to add the second number to the product; finally, to multiply this last result by 10 and to the product add the third number; then desire him to subtract 250 and name the remainder, and you will be able to say what numbers he thought of, and the order in which they were thought of.

The three digits composing this remainder will be the numbers thought of, and moreover they will be placed in the order in which they were thought of. The reason of this will become apparent when we carefully examine the foregoing operations. The person thinking of the numbers, multiplied the first by 2 and added 5, after which he multiplied by 5 and added the second number, by which means his result consisted of 10 times the first number, plus 25; after this he multiplied by 10 and added the third number, so that the result now consists of 100 times the first, 10 times the second, the third, plus 250; then after subtracting the 250, there remains 100 times the first, 10 times the second, and one times the third. Now as each number thought of is less than 10, these numbers must occupy respectively the places of hundreds, tens, and units, which explains the whole mystery of the puzzle.

A perfect number is one such that the sum of all its divisors, except the number itself, is equal to the number.

The following is a list of all the perfect numbers as yet found.

6
28
496
8128
33550336
8589869056
137438691328
2305843008139952128

By inspecting these numbers it will be seen that they all terminate either with 6 or 28.

Amicable numbers consist of two numbers such that the sum of all the divisors of the one is equal to the other number. Thus 220 and 284 are amicable numbers. The sum of all the divisors of 220, including 1, and excluding 220, is equal to 284; and the sum of all the divisors of 284, including 1, and excluding 284, is 220.

The numbers 17296 and 18416 are amicable numbers; so also are 9363584 and 9437056.

Much time and ingenuity have been spent in constructing *Magic Squares*, which consists in arranging the natural numbers up to a certain extent in the form of a square, so that the sum of all the numbers in any row, whether horizontal, vertical, or diagonal, shall be the same. The following are magic squares, the one formed of all the digits, the other all the numbers as far as 16.

2	9	4
7	5	3
6	1	8

1	15	14	4
12	6	7	9
8	10	11	5
13	3	2	16

The sixteen numbers, some of which are negative, when arranged as in the adjoining square, possess some very curious properties.

48	1	21	−22
6	−13	33	44
23	36	−26	27
19	−42	−32	9

The sum of the squares of any four in a row, whether taken horizontally, vertically, or diagonally, makes in all cases the same sum of 3230. This property would claim for it the title of *Magic Square of Squares.* But it possesses the following additional properties: If we fix our attention upon any two parallel rows, whether horizontal or vertical, and take the sum of the products of the corresponding terms, it will equal zero, or in other words, the sum of the positive products will equal the sum of the negative products. Thus, taking the first and second of the horizontal rows, we have

$$48\times 6+1\times -13+21\times 33+-22\times 44=0;$$

and in a similar manner may this property be shown true for other parallel rows.

Three men, Mr. Adams, Mr. Brown, and Mr. Clark, with their sons Daniel, Edward, and Francis, have each a piece of land in the form of a square. Mr. Adams' piece was 23 rods longer on each side than Edward's, and Mr. Brown's piece was 11 rods longer on each side than Daniel's. Each man possessed 63 square rods of land more than his son. Which of the persons were father and son respectively?

Since each man possessed 63 square rods of land more than his son, three distinct sets of numbers must be found such that the difference of their squares shall be 63. And since the difference of the squares of two numbers is the same as the product of their sum into their difference, it follows that 63 must be divided into two factors in three distinct ways. These factors are

$$63=63\times 1=21\times 3=9\times 7.$$

Hence, if we take 63 for the sum, and 1 for the difference of two numbers, these numbers will be 32 and 31. Again, taking 21 for the sum and 3 for the difference of two numbers, we find 12 and 9 for the numbers. If 9 is the sum and 7 the difference of two numbers, these numbers will be 8 and 1. From this we learn that the squares possessed by Mr. Adams, Mr. Brown, and Mr. Clark, were 32, 12, and 8 rods on a side.

Now, since a side of Mr. Adams' square was 23 rods more than a side of Edward's, the square whose side is 32 rods belonged to Mr. Adams, and the one of 9 rods on a side belonged to Edward. Again, since the side of Mr. Brown's square was 11 rods more than a side of Daniel's, it follows that the square whose side was 12 rods belonged to Mr. Brown, while Daniel's square was only 1 rod on a side.

Consequently Francis was Mr. Adams' son, Edward was Mr. Brown's son, and Daniel was Mr. Clark's son.